P9-BZB-555

BODIES OF EVIDENCE

BODIES OF EVIDENCE

BRIAN INNES

The Reader's Digest Association, Inc.
Pleasantville, New York/Montreal

A READER'S DIGEST BOOK

This edition published by The Reader's Digest Association by arrangement with Amber Books Ltd

Editorial and design by
Amber Books Ltd
Bradley's Close
74–77 White Lion Street
London N1 9PF

Project Editor: Brian Burns
Design: Ruth Shane
Picture Research: Lisa Wren

READER'S DIGEST PROJECT STAFF

Editorial Director: Fred DuBose
Project Editor: Nancy Shuker
Senior Design Director: Elizabeth L. Tunnicliffe
Senior Designer: Judith Carmel
Editorial Manager: Christine R. Guido

READER'S DIGEST ILLUSTRATED REFERENCE BOOKS

Editor-in-Chief: Christopher Cavanaugh
Art Director: Joan Mazzeo
Director, Trade Publishing: Christopher T. Reggio

Library of Congress Cataloging in Publication Data
Innes, Brian.
Bodies of evidence : the fascinating world of forensic science and how it helped solve more than 100 true crimes / Brian Innes.
p. cm.
Includes index.
ISBN 0-7621-0295-0
1. Crime laboratories--Case studies. 2. Criminal investigation--Case studies. 3. Chemistry, Forensic--Case studies. 4. Criminals--Identification--Case studies. 5. Crime laboratories--Great Britain--Case studies. 6. Criminal investigation--Great Britain--Case studies. I. Innes, Brian. II. Title.

HV8073 .1444 2000
363.25—dc21 00-031084

Printed in Singapore

Contents

INTRODUCTION

WITHOUT SOME ELEMENT OF FORENSIC SCIENCE, few modern crimes would be solved. If the perpetrator is not observed in the commission of the deed, or if a suspect does not confess, some form of evidence must be obtained, and its validity established in such a way as to secure conviction. In court, expert witnesses will be required to present this evidence, and explain its significance to the jury. Any possibility that the evidence is insecure will be seized on by the defence, and may result in a verdict of Not Guilty. Only the rigour of intensive scientific investigation can ensure that this does not occur.

The word "forensic" means no more than "connected with the courtroom". In the early days of forensic science, almost all those who gave expert evidence were qualified medical practitioners, and well into the 20th century the subject was alternatively referred to as "medical jurisprudence". There was good reason for this: much of the evidence still presented in cases of unnatural death derives in the first instance from the autopsy carried out by a pathologist, or medical examiner. The expertise of specialist toxicologists, serologists, and ballistics examiners, among others, may later be called upon, but it is the pathologist at autopsy who determines the probable cause of death, and provides the samples of tissue and body fluids, whole organs – and even, in most shooting cases, the significant bullet.

In fact, many of the earlier forensic pathologists made important contributions to the development of other branches of the science. They did not confine themselves to post mortem dissection, but also examined trace evidence – both on the body and at the scene of the crime – made deductions from what they discovered, and frequently provided the only evidence necessary to present a cast-iron case in court. It is relatively recently that the great advances in the physical, chemical, and biological sciences have resulted in the establishment of specialist forensic laboratories devoted to the investigation of crime, and the consequent proliferation of experts in specific disciplines.

The earliest known treatise on forensic medicine is the 13th-century Chinese book *Hsi Yuan Lu* (*The Washing Away of Wrongs*). Above all else, this work stressed the importance of examining the scene of crime, stating: "The difference of a hair is of the difference of a thousand li" – a li being a Chinese mile. This adage reflects the importance placed upon trace evidence by the French criminologist Edmond Locard early in the 20th century, an importance recognized by all scene-of-crime examiners today.

In Europe. forensic science developed very slowly. In 1533, the Caroline Code published by the German emperor Charles V was the first to lay down that expert medical testimony must be obtained in cases of suspected murder, wounding, poisoning, hanging, drowning, infanticide and abortion. For some time thereafter, physicians were inhibited by the widely-held objection to dissection of corpses, but this was gradually overcome. The 16th-century French surgeon Ambroise Paré (d.1590) was the first to trace bullets in gunshot victims. In 18th-century Italy, Giovanni Morgagni is credited with the establishment of modern morbid anatomy.

For the present-day reader, the investigations of Sherlock Holmes, detailed in the fiction of Scottish physician Arthur Conan Doyle, are likely to be the first indication of the modern techniques of forensic science, but Doyle was drawing on a knowledge of many established cases. During the 19th century experimental science had made remarkable advances, and police in many countries were quick to exploit its many discoveries. Criminologist Hans Gross first published his *Criminal Investigation* in 1893. In Lausanne, Switzerland, R.A.Reiss established the Institute of Police Science early in the 1900s, and developed forensic photography. Rocard set up his Institute of Criminalistics in Lyon in 1910, and Robert Heindl opened a laboratory, soon to become the German national police laboratory, in Dresden in 1915. The establishment of similar laboratories in Austria, Sweden, Finland, and Holland soon followed.

In the English-speaking countries, development was somewhat slower. The Los Angeles forensic science laboratory dates from 1923, but the FBI laboratory was not established until 1932. In Britain, much early forensic investigation was the province of university departments of medicine, and London's Metropolitan Police Laboratory, under the auspices of the Home Office, was not opened until 1935.

Today, nearly every developed country supports national or regional crime laboratories. The major exception, strangely enough, is the United States. The FBI laboratory is directly concerned only with crimes against Federal law, and cannot apply its formidable expertise except on request from a local police authority. State crime laboratories are well established, and the Medical Examiner system is rapidly spreading, but in many counties the determination of cause of death still remains the duty of the local coroner – an elected office, which may well be occupied by the community's funeral director, without any medical knowledge.

A final word should be said about the use of computers in the solving of crime. They are a formidable tool in the collation of information and the identification of prior offenders. The FBI has its "Big Floyd", but the prize for names must go to Britain's Home Office. In 1987, they announced the setting up of a major system to take over from the Police National Computer. In an obvious tribute to Conan Doyle they named it the Home Office Large Major Enquiry System – to be known to everyone as HOLMES.

Gathering the Evidence

Every contact leaves a trace – every criminal brings something to the scene of the crime, and takes something away. Even a single human hair has a detailed story to tell the forensic scientist back in the laboratory, and it may be the piece of evidence that will complete the case.

MOST MAJOR CRIMES – or at least the more visible ones: murder, assault, rape, kidnapping, arson, explosion, burglary and mugging – occur at a specific time, and a specific place. One can consider all these, loosely, as crimes against the person. There are other crimes – often equally serious – where the criminal activity can be spread over a considerable period, is not directed at any particular person, and may not occur in any specific place. The term "white collar" is often used to describe this type of crime, and is therefore taken to cover topics such as forgery, fraud, embezzlement, and the rapidly growing problem of computer crime.

The investigation and prosecution of almost any crime is likely to require the assistance of a forensic scientist. Forensic experts do not deal exclusively with major crimes: roughly half the work of a forensic laboratory may be devoted to drunk-driving offences and road accidents, and another significant proportion to such matters as drug investigations or industrial accidents. This book, however, concerns itself with the investigation of major crimes, principally with those that take place at a specific location: the "scene of the crime". This is where the majority of the clues to the cause – and the identity of the perpetrator – are likely to be found.

The basic principle of crime scene investigation was advanced early in

CRIME FILE:

Emile Gourbin

"Every contact leaves a trace" was the maxim of French criminologist Dr. Edmond Locard, and he triumphantly established this principle in a sad case of murder in 1912.

Edmond Locard resigned from his post as Professor of Forensic Medicine at Lyon University in 1910, to set up one of the first police laboratories. He put his theory of crime scene investigation into practice in the case of Emile Gourbin in 1912. Gourbin, a bank clerk in Lyon, was accused of the murder of his mistress by strangling, but had an apparently unshakeable alibi. Locard took scrapings from beneath the fingernails of the accused, and examined them under a microscope.

He found flakes of skin that could have come from the victim's neck – at that time there was no way of confirming this but, significantly, they were coated with the same kind of pink face powder that she used. Confronted with this evidence, Gourbin made a full confession, and was subsequently convicted of murder.

CASE CLOSED

the 20th century by Frenchman Dr. Edmond Locard. It is, quite simply, "every contact leaves a trace". In other words, every criminal leaves something at the scene of a crime, and carries something away.

Every inch of a scene of crime must be examined for the tiniest scrap of evidential material. Working shoulder to shoulder, police carry out a 'fingertip search' in a marshy area.

At the Scene of the Crime

It is vital that a crime scene be sealed off without any delay to preserve any evidential trace. This is frequently very difficult: in the case of a suspicious death, for instance, the scene will inevitably be disturbed by the person who found the body, by the first uniformed officers – who are very seldom experts in crime scene investigation – who arrive, by an ambulance crew, and by the medical examiner who pronounces that the body is dead. If the body is out of doors, there will be, at the very least, a profusion of footprints unconnected with the crime. Indoors, the person who first found the body may well have moved it, and loosened clothing – or even removed an object such as a belt or rope around the neck – in an attempt to apply artificial respiration. Significant items in the room may be

displaced. All this will have happened before the scene-of-crime officer ("SOCO", as he is known in Britain) turns up.

The cardinal rule for an investigating officer at the scene of a crime is: "eyes open, mouth shut, hands in pockets". He (it is more likely to be a man) should try to take in every detail: the weather (whether inside or out), the position of the body (whether still alive or dead), and the location of everything that may provide an indication of what has occurred. He should avoid making any comment that could affect the testimony of anybody nearby who may later be called upon to give evidence. He should not touch a thing until his search team arrives. Above all, if there is a hand gun at the scene, he must never – whatever film or television drama has suggested – insert a pencil in the barrel to lift the gun and sniff at it.

These are, of course, ideal requirements. In practice, the investigating officer may well be working on his own during the first, crucial, hours. Weather conditions often make it essential to gather as much evidential material as possible, as quickly as possible. The search team is most probably made up from uniformed officers currently on duty, with little specific training in crime scene investigation. Frequently, even these officers will not be available for some time.

The job of the search team is rather like that of workers on an archeological site – and archeologically trained personnel have sometimes proved the most valuable. Basically, they are looking for something that *should not be there.* It may be a shoe print that does not match that of anybody known to have been present at the site, or signs of a struggle; tracks from a car tire, a fleck of paint caught on a protruding twig, a new scratch on a tree trunk, even a fragment of glass from a broken rear light. There may be fibres torn from clothing; possibly an object that could have been used as a weapon of assault, or an identifiable weapon discarded or hidden some distance from the scene. Other trace evidence will be more obvious: possibly blood – the pattern in which it has fallen is important; cartridge cases from a gun, or bullets that missed their mark. The search team will – hopefully – discover all these.

At the scene of a murder in the Bronx, New York City. Taking care not to disturb anything, police experts examine and photograph everything, before bagging and removing all relevant material for subsequent investigation.

Indoors, there is likely to be other evidence. The team must look for – or discount – signs of forced entry. Overturned furniture, or broken objects, can be evidence of a struggle and, in a case of murder or brutal assault, the perpetrator may try to make the scene look like an attempted burglary. The pattern

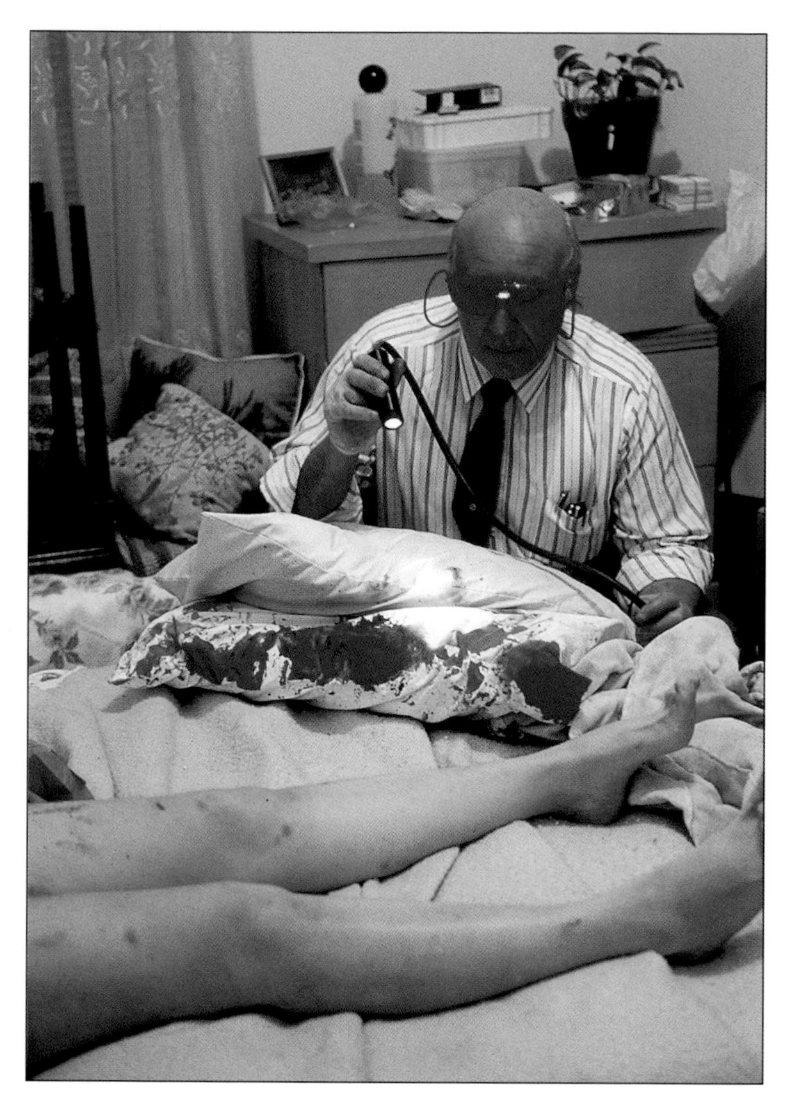

A forensic examiner at the scene of the murder of a 49-year-old woman by multiple stab wounds, in Flushing, New York. Suspecting that rape also occurred, he is searching the immediate area of the bed for semen stains.

of any spilt blood will be more easily determined than in the open air, and can provide vital evidence of the sequence of events (see "Written in Blood").

Investigators at a crime scene must collect each physical object, either between latex-gloved fingers or with forceps, and place it in a plastic bag or box. They must then label this with full details of the time and location, and the precise position in which it was found. Photographs are taken as required, and nowadays a video record is often made of the progress of the investigation. Out of doors, both photographs and plaster casts are taken of shoe prints and tire tracks. Finally, the hands and feet of a dead person are enclosed in plastic or paper bags before the body is removed.

After this initial whirl of activity, the gathering of further evidence can proceed at a more leisurely pace. Indoors, the entire premises must be searched for places where something relevant may be hidden. One can look for fingerprints several hours after the first search, as they are likely to survive a considerable time. Hand prints, and even ear prints – against a window, for instance – are becoming increasingly identifiable. Bloodstains must be scraped up for later analysis. Dust and fibres are collected with a miniature vacuum cleaner. Any documents that may be relevant are collected, as well as ashes from any that have been burnt.

There are two kinds of evidential material. One type is individual, and unique to the crime: pieces of a broken object, tool marks, bullets, or fingerprints, for example. The other is identifiable, but not unique: fibres from a piece of clothing, fragments of paint or glass, etc. The latter are valuable in building a case, and may lead to the criminal, but do not provide proof. Whatever the nature of the evidence, however, it is essential that the "chain of custody" is recorded. Different items of evidence may pass from hand to hand, from one police officer to another, and on to various experts for laboratory examination. Each move must be

CRIME FILE:

Malcolm Fairley

A criminal will often discard vital evidence as he or she leaves the scene of a crime. A painstaking police search uncovered a wealth of trace material that led to the arrest of masked rapist the "Fox".

His head covered with a blanket, Malcolm Fairley, the Fox, is led from court after receiving six life sentences for rape.

The investigation that led to the eventual capture of the brutal rapist the "Fox" in southeast England highlighted the importance of meticulous crime scene searches. During the summer of 1984, the residents around Leighton Buzzard, Bedfordshire, were terrorized by a masked man, armed with a sawn-off shotgun, who broke into houses at dead of night, tied up the men and raped their wives. Several victims reported that he wore his watch on his right wrist – a sign of a left-hander.

On August 16, the Fox struck again. After he had satisfied his lust, he took a hairbrush and carefully combed through his victim's body hair to remove any traces of his own. Then, with a sharp knife, he cut out a large square of the semen-stained bed sheet and, with the knife, brush, and piece of sheet in his pocket, he escaped.

In the morning, police followed the Fox's tracks to the spot where he had left his car. Along the trail they found his shotgun, newly buried in a plastic bag. Only 300 yards (270 metres) from his victim's house, they found the hairbrush and sheet. They discovered footprints and tire tracks where the car had been parked. They found the mask and a single glove half-hidden among the rubbish at the edge of the road,

The glove had a lining of rabbit skin that matched little pieces of fur found at the home of the Fox's first victim, and shreds that had adhered to the material he had used to bind another victim. The mask had been made from the leg of a pair of blue overalls. Finally, the investigators found tiny flecks of paint on a broken sapling where the car had been parked. Laboratory tests identified them as a car paint known as "harvest yellow", a paint that was used only by the British Leyland car company.

A truck driver reported having seen a car backing off the road into the woods at the very spot. Unfortunately, he could not remember the make or colour. Under hypnosis, however, he recalled a harvest yellow Austin Allegro (manufactured by British Leyland) with a Durham registration.

The police now knew a great deal about the Fox, but not who he was. They checked hundreds of suspects, and asked social workers and doctors for the names of any man who had moved recently into the area. One doctor named a Malcolm Fairley, who had arrived from Sunderland, then moved again to north London. Two constables were sent to question Fairley, and found him cleaning a harvest yellow Austin Allegro. His watch was on the dashboard: when he was asked to put it on, he strapped it to his right wrist. In the trunk of the car was a pair of blue overalls with one leg missing. The Fox had been run to earth.

CASE CLOSED

logged and signed for. If this is not done, the defence may justifiably question the validity of the evidence.

The trial of former football star Orenthal James (O.J.) Simpson, for the murder of his estranged wife Nicole and waiter Ronald Goldman on June 12, 1994, revealed how a crime scene investigation can be mishandled, and the subsequent chain-of-custody requirements ignored.

First of all, police kept the bodies of the victims lying in the open for more than ten hours – covered with a blanket taken from Nicole Simpson's home! – before a medical examiner was allowed at the scene. And then, at the trial, the forensic pathologist who performed the autopsy admitted that he had made up to forty errors during examinations.

What appeared to be irrefutable forensic evidence included blood splashes found at the scene and typed as matching Simpson's blood; a pair of socks soaked with blood, found at the foot of his bed, that matched that of the victims; and a similarly blood-stained glove, allegedly found behind his house, that matched one found at the murder scene.

However, it emerged in court that a phial containing a sample of blood taken from Simpson had mysteriously decreased in volume by 1.5 ml while in police custody – which immediately raised a suspicion that evidence might have been planted. Two defence experts who examined the socks two weeks after the murders testified that they saw no signs of bloodstains – and the prosecution had to admit that the stains were not discovered and reported until four weeks later. Samples extracted from the socks were sent to the FBI laboratory in Washington DC. They were found to contain EDTA, a preservative added to blood samples to prevent coagulation. And as for the gloves – they were apparently too tight for Simpson's hands.

DNA evidence was presented in a confused and confusing way, and it is unlikely that the jury appreciated its true significance. And when the officer who found the glove, Detective Mark Fuhrman, admitted

June 12, 1994: The body of Nicole, wife of O.J. Simpson, lies in a pool of blood at the foot of the steps leading to her home in the Brentwood section of Los Angeles. Bloody prints stain the path.

perjuring himself in earlier evidence, the prosecution's case fell apart.

The jury acquitted Simpson on September 30, 1996, after three hours of deliberation. However, in a civil case for wrongful death subsequently brought by Ronald Goldman's father, he was found guilty of both murders.

If the crime has involved fire or explosion, preliminary examination of the scene is unlikely to yield much in the way of useful evidence, and the specialist experience of fire officers or explosives experts will be necessary. In the case of an aircraft crash, for instance, where there are a number of dismembered bodies, the problem of identifying the victims, and re-assembling their remains, requires the assistance of experts in forensic anthropology and odontology (those with particular experience in the subject of teeth).

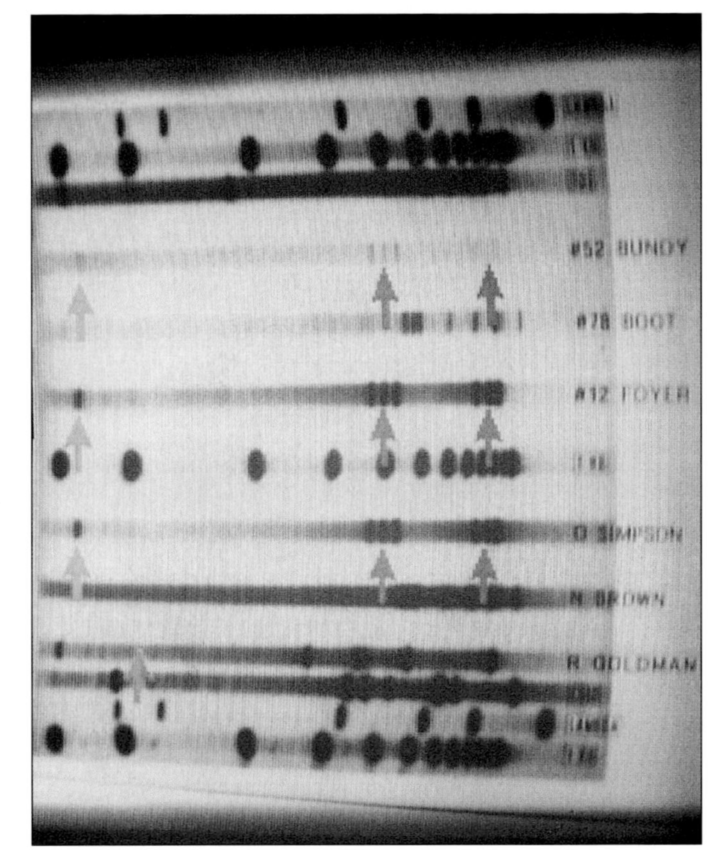

During the trial of O.J. Simpson, monitors in the courtroom were used to display the DNA evidence. Arrows indicate the matching DNA bands from samples taken at the scene of the murders, the hallway of his home, and from Simpson himself.

In the Autopsy Room

The task of gathering evidence continues in the autopsy room (or, in the case of assault or rape, with a full physical examination of the victim).

The word "autopsy" means "seeing for oneself", and this is exactly what the pathologist sets out to do. He or she has the task of examining a dead body in detail, and, if possible, determining the cause of death. Clues to the identity of the victim may also be necessary.

Firstly, the examiner must ascertain that the victim is definitely dead. There have been many unfortunate occasions when the first person to examine the body has pronounced that death has occurred, but the "corpse" has subsequently shown signs of life in the mortuary, or even on the dissection table. Drug overdoses, other forms of poisoning, or electrocution, can induce a state of "suspended animation": there is no discernible heartbeat or respiration, and even the electrical activity of the brain may not be detected, but the victim can subsequently be revived in the intensive care unit.

Establishing the time of death is very important, particularly if suspects will later need to provide alibis. Unfortunately, although claims have been made for the relative accuracy of certain techniques, no method can provide more than a rough estimate. One can determine the precise time of death in only a few cases, such as when a clock has been stopped by a bullet.

CRIME FILE:

Sidney Fox

Sidney Fox, who murdered his mother in a Margate hotel in October 1929.

Even forensic experts can disagree. The question of whether or not a bruise had been found in Mrs. Fox's larynx was hotly debated in court, but in spite of the doubt the jury found her son guilty of murder.

Sidney Fox and his mother Rosaline booked into the Metropole Hotel, in Margate, southeast England, on October 23, 1929. At 11.30 PM, Sidney Fox raised a cry of "Fire!", and Mrs. Fox was found dead in a smoke-filled room, where an armchair lay smouldering. Two doctors who were summoned both agreed that she had died of shock, a verdict that was confirmed at the coroner's inquest the following day.

Fox, however, had renewed his mother's life insurance for a single day on

The open doorway leading to the adjoining bedroom, which was occupied by Mrs Fox's son.

October 22. The insurance company became suspicious, and informed the police. When Mrs. Fox's recently buried body was exhumed, the distinguished Home Office pathologist, Sir Bernard Spilsbury, examined it. He found nothing to account for heart failure due to shock – although he detected progressive disease in the heart and arteries – nor any signs of asphyxia due to inhalation of smoke. What he did find, he later testified at the trial of Fox for murder, was a circular bruise in the soft tissue between Mrs. Fox's larynx and oesophagus. It was about "the size of a half-crown" (about $1\frac{1}{2}$ inches, or 3 centimetres across), and he deduced that Sidney had strangled his mother while she was asleep, before starting the fire.

The defence called two expert witnesses, one being the equally distinguished pathologist from Edinburgh University, Sir Sydney Smith, and the other Dr. Robert Bronté. Both men had viewed Mrs. Fox's larynx, and found "putrefactive discoloration", but they agreed that there was no sign of a bruise. Spilsbury assured them that he had seen it at the time of the exhumation, but that it had become obscure before he was able to take a section for microscope examination.

Smith later wrote in his autobiography, *Mostly Murder*: "A microscopical section would have been of inestimable value in showing whether the patch of discoloration ... was a bruise or not. Personally I was pretty sure it was not."

Smith was aggressively cross-examined in court on the question of whether a pathologist could distinguish between bruises and local discoloration. "Do you suggest Sir Bernard would not know the difference between the two?" he was asked. He replied that nobody could tell only by sight, and added: "I do not think that anybody should say a bruise is a bruise until it has been proved that it is."

There was also dispute over the fact that the tiny hyoid bone in Mrs. Fox's larynx, which is easily broken in cases of strangulation, was intact. In his summing-up, Mr. Justice Rowlatt said that this fact was "a very strong point in favour of the accused". Nevertheless, Sidney Fox, protesting his innocence to the end, was found guilty of his mother's murder, and was hanged on April 8, 1930.

Another view of the scorched armchair in front of the gas fire. It is evident that this could have produced only a limited amount of smoke. The singed section of carpet shows where it originally lay on its side.

It was once customary for the first doctor at the scene of the crime, after he or she had established that death had occurred, to take the temperature of the body, usually by a thermometer deep in the rectum. This, however, usually results in disarrangement of the clothing, and inevitably interferes with the pathologist's examination for semen, blood, hairs, and other evidence. It is safer to leave such an estimation of time of death until after this examination had been completed.

The body begins to lose heat from the moment of death. A clothed body of average build, in temperate regions, shows a fall in temperature of about 1.8°C (3.2°F) per hour over the first six to eight hours, after which the rate of cooling decreases. Unclothed bodies cool quicker, while fat bodies cool less rapidly. The rate of temperature loss is similarly related to the ambient temperature: in very hot climates, for example, there may be no cooling at all – the body may even become warmer after death.

All these considerations – and many others besides – must be taken into account. One generally assumes that the body temperature was 37°C (98.4°F) at the moment of death, but people who die from hypothermia, for instance, will begin cooling from a markedly lower temperature. Complicated graphs and formulae have been drawn up to allow for all sorts of conditions. Even the best of these, however, can claim an accuracy no greater than 2.8 hours each side of the estimated time.

During the autopsy, samples are taken of various body fluids, including the blood, urine, and the liquid that surrounds the brain. It has been claimed that changes in the chemical constitution of these fluids can provide an indication of the time of death, but there is no way of allowing for physical or emotional conditions that may affect the rate of change. Another proposed technique is the analysis of the potassium content of the vitreous humour of the eye, which increases steadily for four or five days following death. However, since nobody knows the initial potassium level of the living eye, this method is no more reliable than any other.

A further indication of time of death is the onset of rigor mortis. In

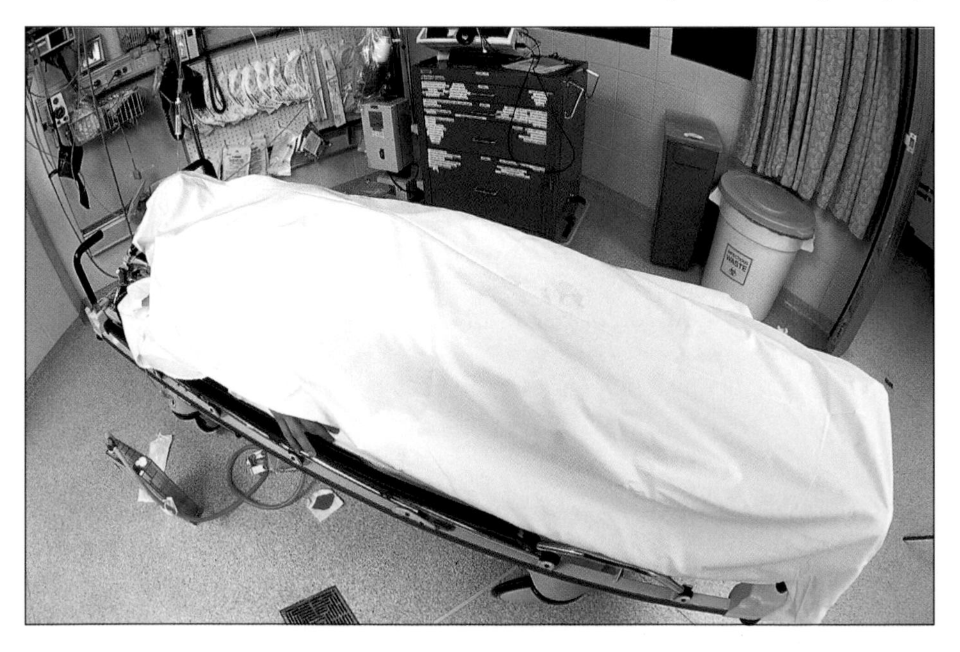

In a corner of the autopsy room, the body of a dead person awaits transfer to a dissection table for post mortem examination.

normal conditions, the muscles of the face begin to stiffen within one to four hours, and the limbs in four to six hours. After 12 hours, the body is rigid, and then gradually relaxes as tissue decomposition sets in. Again, these changes are subject to wide variations.

Rarely, and generally only in conditions of extreme emotion or violence, rigor mortis may set in immediately after death. A soldier, killed by a shell at the siege of Balaclava, Crimea, in 1854, is said to have stayed mounted upright on his horse. At the defence of Sedan during the Franco-Prussian war in 1870, the decapitated body of a soldier remained sitting rigidly, his hand firmly grasping a cup.

During the autopsy, the pathologist talks through every stage of the examination. In the past, this required a note-taker, but nowadays a tape recorder is usually used. Photographs are taken of any significant details revealed during the autopsy, and a video recording may be made of the entire procedure.

First, the pathologist describes the outward appearance of the corpse: physical features, racial type, and any clothing, which may be visibly damaged by a weapon. After the clothing has been carefully removed, or cut away if necessary, the external condition of the body is examined closely. The colour is important, as it may indicate poisoning, particularly by carbon monoxide. All contusions and wounds must be described, together with the condition of the eyes. The examiner will also look for hypostasis.

Hypostasis, or post-mortem lividity, sets in immediately after death. When the heart stops, the circulation ceases at once, and gravity causes the blood to sink through the blood vessels to the parts of the body that are lowest. The red corpuscles settle first, and become visible as pinkish-blue patches about one to three hours after death. After six to eight hours, these patches join up into purple-red areas. They do not form where the weight of the body pressing against a hard surface prevents the accumulation of blood. In the case of a body lying on its back, for example, they are found on the back of the neck, the small of the back, and the thighs; while a hanged body will develop hypostasis in the hands and legs.

The appearance of these dark patches can be a useful indication that a body has been moved to a different position some hours after death. Sometimes, the first officers on the scene have mistaken them for bruises, and assumed that the body has been badly beaten. Although the pathologist can soon determine the true nature of any discoloured patches, even experts have been known to disagree about bruises.

If the identity of the victim is unknown, fingerprints are often taken during the preliminary examination. The investigating officer examines the victim's clothing for any clues. If there are no documents – credit cards, driver's licence, letters, bills, even theatre or cinema ticket stubs – that can help to establish identity, manufacturers' labels in the clothing and shoes must be

CRIME FILE: Patrick Higgins

The condition of the contents of the stomach can sometimes be an indication of the time of death. Undigested vegetables from a Scotch broth showed that two little boys had been drowned shortly after eating it.

One summer Sunday afternoon in 1913, two men noticed a dark bundle floating in a flooded quarry in West Lothian, Scotland. To their horror, they discovered that it was two small bodies, tied together with cord.

Forensic expert Sydney Smith, from Edinburgh University, examined the bodies. He soon determined that they were those of two young boys, whose ages he estimated at seven and four. The remains of their clothing were similar, and of identical make, suggesting that they were brothers. One of their shirts carried a faint laundry mark from a poorhouse in Dysart, Fife.

Because the boys had been immersed in water, their body fat had been converted into a substance known as adipocere (see "Breath of Life"). This transformation had prevented their bodies' decomposition – leaving the stomachs intact, and their contents almost unchanged. Smith discovered that: "In each stomach were several ounces of undigested vegetable matter – whole green peas, barley, potatoes, turnips and leeks: in fact the traditional ingredients of Scotch broth."

From the condition of the adipocere, Smith calculated that the boys had been in the water between eighteen months and two years, and that they had therefore taken their last meal in the summer or autumn of 1911.

Inquiries managed to establish that two local boys, aged seven and four, had disappeared in November 1911, and had previously been in the poorhouse in Dysart. Their father, a labourer named Patrick Higgins, had told an acquaintance in November: "The kids are alright now. They're on their way to Canada." When police discovered a woman in the area who remembered giving two boys a meal of broth one November night, the case against Higgins was established. He was arrested at his lodgings, found guilty at trial, and hanged in Edinburgh in October 1913.

CASE CLOSED

noted. A cast or print of the shoes should also be taken, to eliminate some of those taken at the scene of crime. Casts of the teeth may be taken at this stage, or later by an expert odontologist. Finally, the body is examined for marks of injections, but these can be very small or inconspicuous. This is particularly true if – as is often the case where the dead person has been a user of drugs – the arms are tattooed.

The pathologist is now ready to begin the autopsy proper. First of all, swab specimens are taken from such sites as the hands, mouth, breasts, vagina and rectum. In a case of sexual assault, the pubic hair is combed and placed in an evidence bag or box, in the possibility of discovering some foreign hair, and the rectum is similarly examined.

The first stage of the internal examination is to make a large Y-shaped incision, which begins behind each ear and extends down the sternum and as far as the groin. This enables the pathologist to peel back the skin, exposing the neck and chest, and revealing the bones, muscles and internal organs. It also uncovers any subcutaneous bruising that may not have been apparent during the external examination. Tissue samples are taken from wounds and contusions, and all wounds are explored and described carefully. In the case of a shooting, the bullet or bullets must be recovered. The pathologist also examines any broken bones – particularly, in a case of strangulation, those of the neck.

The breast bone must be cut through, to remove the lungs, heart, and other organs, which will be required for subsequent examination and analysis. Then the skull must be opened. The initial incision is continued across the top of the head, and the skin is pulled away to expose the bone. A circular saw is used to cut round the skull, and the top is prised off. The pathologist should examine the brain and the inside of the skull closely for any injury – traces of old injuries may give a clue to the victim's previous lifestyle. Then the brain is removed for later examination. An experienced pathologist can complete the entire operation in half an hour or less – the distinguished English pathologist Francis Camps claimed that he could often perform it in ten minutes.

In the case of a body associated with a fire, or apparently drowned, the pathologist searches the air passages for traces of soot or water. He or she also examines the stomach contents, as they can supply evidence of the time elapsed between the last meal and the hour of death (see Crime File, opposite).

During the autopsy, the pathologist sometimes detects signs that suggest poisoning. Here the sense of smell is very often important – to detect substances such as ammonia or phenol (carbolic acid) in the stomach, or the characteristic bitter almond odour of cyanide. In cases of suicide, all sorts of chemicals may have been taken. In an unusual case, a young girl was found dead in bed. On opening the skull, her brain smelt of the cleaning fluid carbon tetrachloride, which she had been in the habit of sniffing.

The condition of the liver may indicate cirrhosis or hepatitis, though many drugs – and particularly an overdose of paracetamol – can produce a similar appearance. Further investigation in the laboratory will be necessary. Inflammation of the kidneys can be due to poisonous metallic salts such as mercury compounds, chronic lead poisoning, or long-term over-use of phenacetin.

The initial work of the pathologist is now complete. Under his or her supervision, assistant specialists in serology and tissue analysis, odontologists, toxicologists, and forensic anthropologists, will usually continue the investigation. Their work is described in later chapters.

Suicide or Murder?

Frequently, when a person is found hanging in what is taken for a suicide, the rope is cut by the first person finding the body, in an attempt to apply artificial respiration. This can destroy vital evidence, such as the attitude and undisturbed position of the body.

IN THE CASE OF A DEATH, ONE MUST ASK: is this natural, accidental, or unnatural? If unnatural, is it suicide or murder? There are many reasons why people commit suicide – and there are many reasons why people commit murder. In some cases, one may very well be taken for the other. As Professor Cyril Polson of Leeds University put it in his *Essentials of Forensic Medicine*: "The cunning suicide … may plan his death in a manner which suggests homicide. More often, attempts are made by a murderer to present the result of his crime as a suicide."

Professor Polson also stressed, in an article in *The Criminologist*, that the apparent cause of death can be very misleading. He described twelve cases – apparently three shootings, two stabbings, two stranglings, attacks with a bottle and an axe, a kicking, a smothering, and a violent beating – which appeared at first sight to the police to be murder. All were later proved to be the result of suicide or accident. The "axe murder" turned out to be a shotgun suicide; one of the "stranglings" was a heart attack; death by "kicking" was caused by an accidental fall – and the "sea of blood" attributed to a violent attack was produced by a burst varicose vein.

The eminent pathologist Sir Sydney Smith also came across some bizarre cases during his career. In *Mostly Murder*, he described how "a maid in a hospital hacked the front of her head, inflicting twenty cuts; then, finding this ineffective, she filled a bath with warm water and drowned herself. I wonder how many of us

The eminent pathologist Sir Sydney Smith. In his distinguished career, Smith came across more than one extraordinary case in which all the initial signs pointed to murder; only after further investigation could a verdict of suicide be safely returned.

doctors, finding a number of hatchet wounds in the skull, would think of suicide."

On another occasion, wrote Smith, a man was found hanging. He had a bullet wound on the right-hand side of his face, and another in the palm of the left hand; five cuts in his throat; and cuts in his left wrist that had severed the muscle tendons but not the main blood vessels. Yet it was clear from the circumstances that the man had committed suicide: he had first attempted to shoot himself, then cut his throat and wrists; finally, in desperation, he had hanged himself.

In his university lectures, Smith would often conclude with a case (not one of his own) in which a man set out to hang himself from the branch of a tree that was on the edge of a cliff, and stretched over the sea. First he took a large dose of opium, then decided to make certain by shooting himself as well. "The noose adjusted, the poison taken, and the revolver cocked, he stepped over the cliff, and as he did so he fired. The jerk of the rope altered his aim, and the bullet missed his head but cut partly through the rope. This broke with the jerk of the body, and he fell fifty feet into the sea below. There he swallowed a quantity of salt water, vomited up the poison, and swam ashore a better and wiser man."

A Curious Case in Edinburgh

One of Sir Sydney Smith's cases highlights the care that one must exercise in investigating what may appear at first to be murder. An elderly man left his hotel in Edinburgh, Scotland, one night, and did not return until 7.30 AM the following morning. A maid who opened the door saw that he had blood on his face, but he said "Don't worry, I will go upstairs and have a wash". He hung up his coat, hat and umbrella, and walked upstairs to the bathroom, where he collapsed. He was rushed to hospital, but died three hours later without regaining consciousness.

It was obvious that he had been shot in the head. The gun had been held beneath his chin and the bullet had gone through his brain, causing extensive damage, and had emerged at the left of the frontal bone of his skull. The exit hole was 1¼ inches (3 centimetres) in diameter, the size and shape suggesting that a .45 revolver bullet had turned on its side before exiting.

CRIME FILE:

Iris Seagar

Did she fall or was she pushed? Ingenious experiments by a forensic scientist clearly established that a Baltimore woman had not fallen or jumped, but had been thrown from a balcony.

In the early 1970s, Iris Seagar plunged 200 feet (61 metres) to her death from the balcony of her penthouse in Baltimore, Maryland. Neighbours suggested that the drunken behaviour of her husband was enough to have driven her to suicide, and the police were prepared to leave it at that. The husband, however, claimed that the death was an accident. "She was fiddling with a faulty air-conditioner," he said, "and tipped over the guard-rail." When the police learnt that Mr. Seagar was the beneficiary of his wife's $100,000 insurance policy, which would have been invalid in the event of suicide, they were forced to investigate further.

A forensic scientist constructed several dummy models of the weight and height of the 48-year-old woman, and a video camera recorded what happened as they were tipped, pushed and thrown from the balcony. The recordings established that, if Mrs. Seagar had fallen accidentally, her body would have landed not more than 10 feet 6 inches (3.2 metres) from the foot of the building. If she had jumped, the distance would have been no more than 14 feet (4.3 metres). In fact her body was found 16 feet (5 metres) from the building. Faced with this evidence, her husband confessed that he had thrown her from the balcony in a drunken rage.

CASE CLOSED

The police followed a trail of blood from the hotel to a shelter in gardens across the street. There they found a .45 revolver and a large pool of blood. There was a bullet-hole, surrounded by bone and brain fragments, in the roof of the shelter above. It had begun to snow at about 6.00 AM, and a single track of footprints and blood spots led from the shelter in a wide circle, and from the shelter to the hotel.

The gun turned out to be the property of the man in question, and letters later discovered made it clear that he had committed suicide. Examination of the blood trail indicated that he had shot himself some time before 6.00 AM. Apparently he had then sat on a seat in the shelter for some time, with his head hanging forward, resulting in the pool of blood. Following this he walked in the gardens before returning to the shelter; after resting again, he walked to his hotel. Despite the fatal damage to his brain he had somehow survived for two or three hours, carried out several purposeful acts, and even spoken clearly before losing consciousness.

Similarly, in 1992 a Dutchman, who had abducted and murdered two young girls near Perpignan in southern France, was captured alive in a hotel in Lourdes: he had gone there to confess his guilt. On finding the town's religious shrine closed for the night, he had attempted first to electrocute himself, then to cut his wrists, and finally, and equally unsuccessfully, to shoot himself.

Another leading pathologist, Dr. Keith Simpson, was called to a bizarre case of suicide in England in 1945. A man's body was found in the water of Portsmouth Dockyard, trussed up with rope. Examination established that the dead man had drowned. Although the police were convinced that it was a case of murder, Simpson disagreed. The man had died, he said, "by his own hands and teeth." He demonstrated that he had tied himself up, beginning with a noose around his lower legs, pulling the rope upward at each knot, and making the final knot tight with his teeth. And, shining a torch into the man's mouth, Simpson pointed out a strand of the rope caught between two teeth.

Some time in the 1930s, a 45-year-old woman was found dead in bed. The bedroom door was locked, and had to be broken down, and the key was found on the inside. The woman lay on her back, covered by the bedclothes but with both arms outside. There was a scarf, folded to a width of 7 inches (10 centimtres) over her mouth, and knotted at the back. Another scarf of about the same width was round her neck, tied in front tightly enough to mark the skin. Finally, a small handkerchief had been forced into the back of her throat.

The fatal wound in a genuine suicide is almost always preceded by one or more tentative attempts.

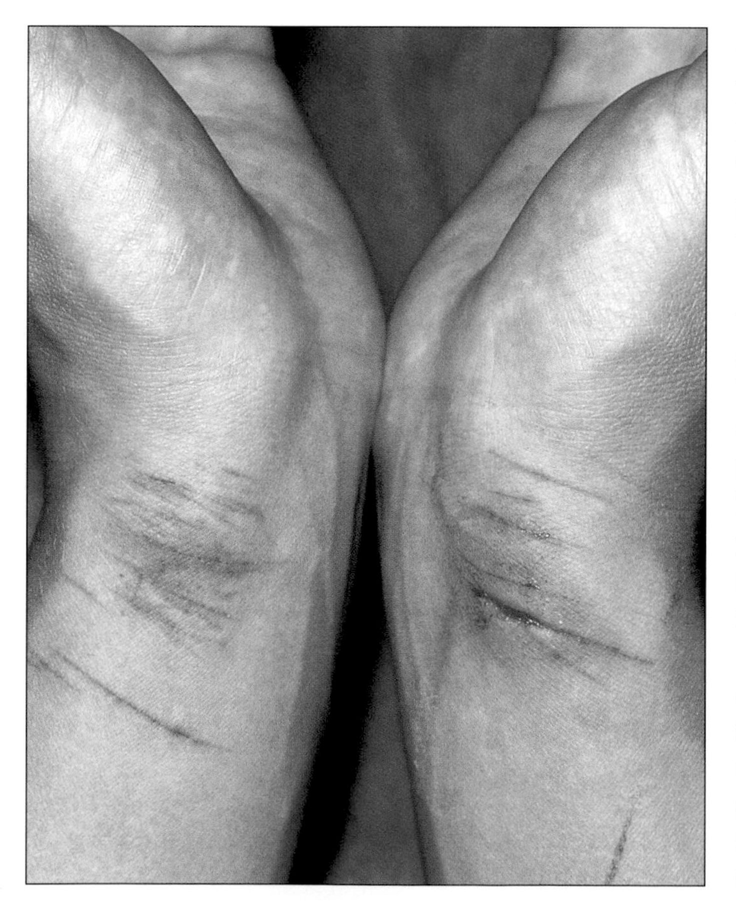

However, inquiries revealed that the woman had suffered from deep depression for several years, and had sometimes threatened to take her life by tying a handkerchief or scarf round her neck. If it had not been for the circumstances in which the body was found, in a locked room, it might have been easy to assume her death to be a case of murder.

Nobody knows how many cases in which the cause of death was deemed to be suicide may actually have been murders. Without an obvious suspect or overt suspicious circumstances, and where there are reasons to believe that the victim was likely to take his or her own life, it is tempting for the police to avoid a great deal of unnecessary investigation. The same is true of deaths that appear to be accidental.

Over the years, there have been many cases of apparent suicide by cutting the throat or wrists. A genuine suicide is almost always preceded by two or three tentative attempts, which leave cuts insufficiently deep to result in death. A single deep cut, therefore, can be a strong indication of

CRIME FILE:

Heinz W.

The taxi driver from Berlin claimed that the German couple had committed suicide with sleeping tablets, but toxicology tests showed that they were only unconscious when he shot them both in the head.

German couple Iris and Johannes Gerarts were found dead in bed at their elegant villa in Kiel on Easter Monday, 1998. Both had gunshot wounds to the head.

The couple were the wealthy owners of a garden design business, and several valuable items had been stolen from their home. In addition, money had been withdrawn from their bank account after their deaths.

Shortly afterwards, police arrested a 50-year-old Berlin taxi driver, Heinz W., and charged him with murder. He had worked as a driver and private detective to the couple, but denied the charge, claiming that he had actually been helping them to commit suicide. The two had been in poor health (Johannes had had one leg amputated as a result of diabetes, while Iris suffered from multiple sclerosis) and their business had been in financial difficulties. The driver claimed they killed themselves by taking an overdose of sleeping tablets, having first asked him to shoot them after they had died and make the suicide look like a robbery.

However, toxicology tests found that although the couple did have traces of sleeping tablets in their bodies, it was nowhere near a fatal dose. Heinz W. then changed his story, admitting that they had taken the tablets so that they would be asleep when he killed them. His version of events stood up, and instead of the five years to life he would have earned for murder, he was sentenced to four years and nine months for assisting a suicide, violations of gun laws, and faking a crime.

CASE CLOSED

homicide made to look like suicide. The same is true in Japan, where ritual disembowelling, although illegal, is still a traditional form of suicide. Colin Wilson, in his *Written in Blood*, reported an observation made by Inspector Asaka Fukuda. During a long police career, he had witnessed a number of cases that proved to be homicide. In nearly all cases of genuine suicide, said Inspector Fukuda, the perpetrator first made one or more trial pricks with his knife or sword before driving it deep into his abdomen. In the absence of this "hesitation injury", Fukuda always looked for further evidence of murder.

There are also cases of murder that have been made to look like suicide by hanging. Often the murderer claims to have found the body suspended, and to have cut it down in an effort to revive it. This conveniently destroys the most valuable piece of evidence: the attitude and undisturbed appearance of the body. Post-mortem examination, however, and a careful search of the scene, will invariably reveal incriminating details.

CRIME FILE:

Henry Marshall

His death was declared a suicide, but there seemed no reason why the US Department of Agriculture official should have killed himself. His involvement in a high-level financial investigation raised suspicions of murder, and the later exhumation of his body established the fact.

Henry Marshall was an official of the US Department of Agriculture, and was investigating a case of misue of funds when he was found shot near Franklin, Texas, on June 3, 1961. An inquest decided he had committed suicide, but lingering doubts resulted in his exhumation on May 22, 1962.

On June 3, 1961, the body of Henry Marshall was found lying beside his pick-up truck at his ranch near Bryan, Texas. There were five bullet entry holes in the front of his body, and four at the back, but no bullet in his body. His .22 rifle lay beside him, with four spent cartridges. Powder burns on the front of the body showed that the rifle had been fired at point-blank range. The local justice of the peace certified the death as suicide, and Marshall was buried.

Marshall had been an official of the US Department of Agriculture, and had been involved in an investigation of serious financial misdemeanour. The following year, a Washington spokesman suggested that Marshall had been murdered, and an order was issued for his body to be exhumed. A state medical examiner, Dr. Joseph Jachimczyk, carried out the autopsy, and found a near-lethal concentration of carbon monoxide – possibly from the truck's exhaust – in Marshall's lungs. As he was unable to explain the absence of a fifth bullet exit hole, Jachimczyk suggested that perhaps two bullets had emerged from the same hole. He also found evidence of a crushing blow to the head.

It seemed impossible that Marshall could have inhaled a large quantity of the truck's exhaust and then shot himself five times with the rifle, one that needed cocking for each shot – particularly as a permanent injury to his right arm made it impossible for him to straighten it completely. A Senate sub-committee and the Texas Rangers agreed that this was a case, not of suicide, but of murder. Unfortunately, the murderer was never found.

Following Henry Marshall's exhumation, further forensic examination established that he had been murdered.

A murderer will always need to use some degree of force, for instance. It would be almost impossible to persuade the victim to cooperate in setting up the scene: to put the noose about their neck and stand on a chair, for example, which could then be kicked away. There is likely to be evidence of a struggle, or marks on the floor where the murderer has dragged the body. If he or she then passes the rope over a beam or hook and hauls it up, the fibres of the rope may well show signs of this hauling. Alternatively, the victim may have been strangled on the ground and the cut portion of the rope subsequently looped over a beam to accord with the murderer's statement. Subsequent examination of the wood will not reveal any of the marks that a heavy weight would have made. Inspection of the cut ends of the rope under the microscope can show that it was not cut under tension – in a genuine suicide, the rope would have been pulled taut by the weight of the body.

Deaths by poisoning have frequently been attributed to suicide or accident, particularly where the victim was a regular taker of drugs, either on prescription or to provide some kind of stimulation. Nobody knows how many doctors have signed a death certificate in such a case, when the drug has in fact been administered by someone else – either to end the suffering of a terminally ill patient, or as an act of murder.

A notorious English case was the death of Charles Bravo on April 21, 1876. Three days earlier, Bravo dined with his wife Florence and her companion Mrs. Jane Cox, and drank a glass or two of burgundy. Late in the evening, he was

The inquest held upon the unexplained death of Charles Bravo in 1876. Inset are portraits of the dead man and his wife, Florence.

Known as hara-kiri *in the West, but as* seppuku *in Japan, ritual self-disembowelment was regularly practised for centuries. Although it has been made illegal, it is still occasionally performed in cases of great dishonour. It was a slow and agonising death, and disgraced samurai were permitted to have a* kai-shaku, *or second, whose task was to strike off the man's head as soon as he made the first cut.*

heard calling from his bedroom, and Jane Cox found him vomiting, after which he lapsed into unconsciousness. When a doctor arrived, Cox informed him that Bravo had told her he had taken poison. When the sick man regained consciousness, he admitted that he might have swallowed some laudanum, which he had rubbed on his gums to ease his neuralgia.

Florence Bravo sent for leading physician Sir William Gull, who told her husband that he was certainly poisoned, and dying. Bravo insisted that he had taken nothing but the laudanum. An analysis of his vomit, however, revealed that he was dying of antimony poisoning, probably in the form of a medicinal remedy known as tartar emetic. There were no traces of antimony in the burgundy he had drunk, however, and his wife and Cox had eaten the same dishes.

Sir William Gull was convinced that "whatever was taken, he took it himself", and the coroner at the subsequent inquest agreed with him. The "Bravo Affair" soon became front-page news and old scandals concerning Florence Bravo emerged. A second inquest opened in July 1876, at which the exhumed body of Charles Bravo was exhibited to the jury. It became virtually a trial of Florence Bravo and Jane Cox, but, although the jury returned a verdict of wilful murder, there was "not sufficient evidence to fix the guilt upon any person or persons".

Did Mrs. Cox or Mrs. Bravo poison Charles Bravo? And if so, how?

Nowadays, the ingenuity of a murderer seldom matches that of the forensic scientist. If there is any reason to believe that a death is neither suicidal nor accidental, meticulous investigation will always reveal those tiny pieces of evidence that will establish the truth. "Murder, tho' it have no tongue, will speak with most miraculous organ."

CRIME FILE:

Norman Thorne

The poultry farmer from Sussex claimed that his lover had hung herself, after she paid him a visit in December 1924, but the forensic evidence showed otherwise. There were no rope marks on the wooden beam, and Elsie Cameron's neck appeared to be undamaged.

Norman Thorne and his lover, Elsie Cameron, whom he murdered on December 5, 1924.

Elsie Cameron, a London typist, disappeared on December 5, 1924, on her way to the Sussex poultry farm of 24-year-old Norman Thorne, her lover. Five days later her father, who had heard no word from her, contacted the police. When police came to the farm, Thorne told them that he too was concerned, as she had never arrived at his home. A month later, when the police learnt that Cameron had been seen on her way to the farm, they paid another visit, and unearthed her suitcase. Thorne then told a different story. Cameron, he said, had in fact arrived, and announced that she was staying until he agreed to marry her. Thorne had gone out and, returning late in the evening, had found her hanging from a beam in a poultry shed. He had panicked, dismembered her body, and buried the pieces in his chicken-run.

Detectives pointed out that there were no deep rope marks on the beam, such as the jerk of a hanging body would have caused, and that the thick coat of dust on its upper surface was undisturbed. Sir Bernard Spilsbury examined the exhumed remains, and found extensive bruising on the head, face, elbows, legs and feet. Dissecting her neck, he found no evidence of hemorrhaging consistent with hanging.

The trial of Norman Thorne provided a dramatic confrontation between two pathologists. The defence called Dr. Robert Bronté, who had carried out a second autopsy on Cameron, at which Spilsbury was present, nearly a month after her body had been buried. Bronté claimed that he found "grooves" in her neck, with visible bruising; Spilsbury denied that this was so.

On the last morning of the trial, Spilsbury made his final point. Thorne had claimed that when he found Cameron hanging, "her eyes were open, but screwed up". Spilsbury said: "Assuming unconsciousness to have intervened, if not death, the eyes ... would not have been completely closed, or completely open; a half-open condition, with flexive lids; certainly no puckering.'" Norman Thorne was found guilty, and hanged on April 22, 1925.

CASE CLOSED

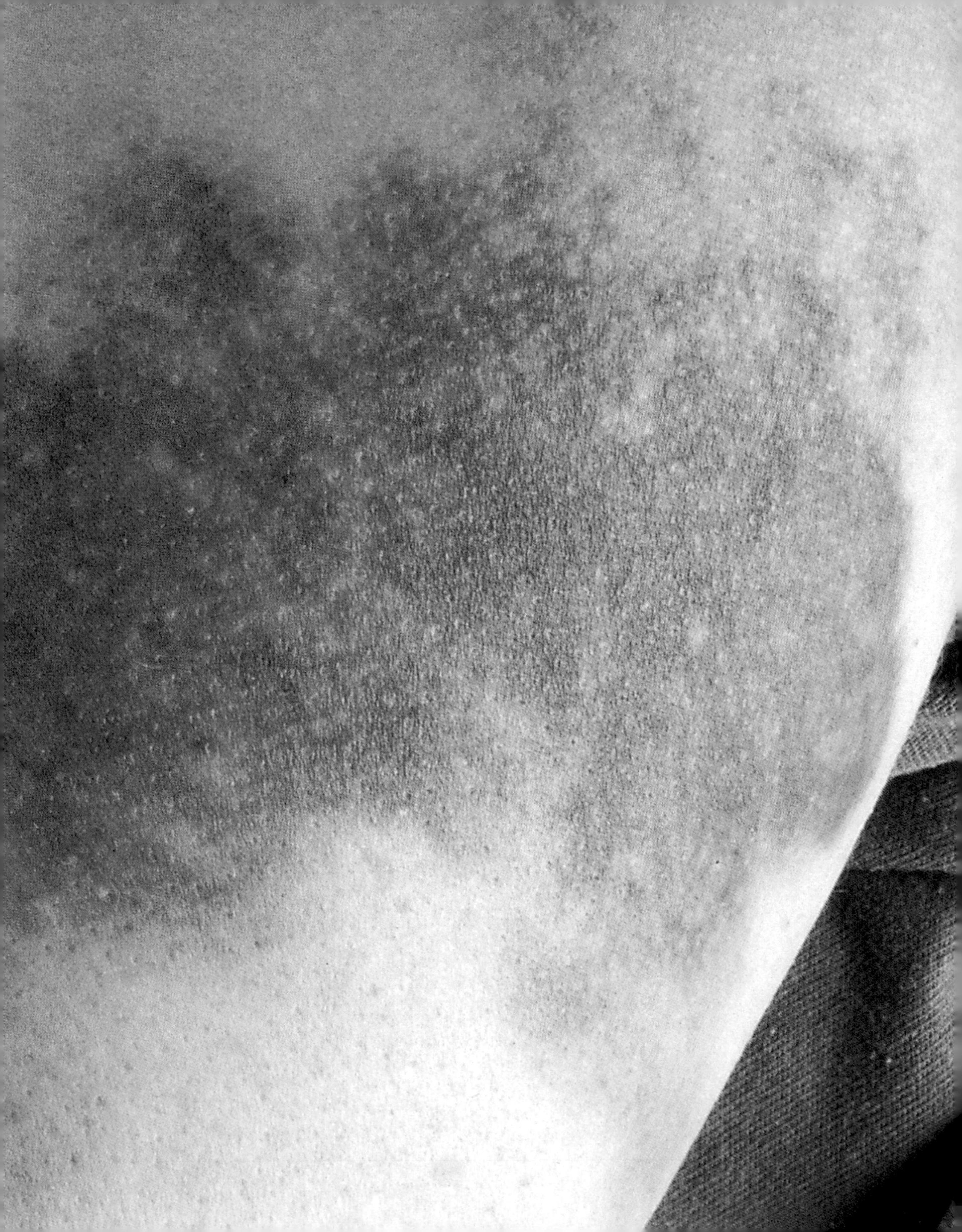

Mark of Death

Severe bruising of a human arm. True bruises can be produced only while the body is still living, being produced by the escape of blood into the surrounding tissues. They will indicate the point of injury, but not often its cause.

PHYSICAL ASSAULT – WHETHER UNPREMEDITATED, or a coldly planned attack – can often lead to death. One can use a knife for stabbing, causing mostly internal injuries and consequent internal loss of blood; or for slashing, when the severing of a major blood vessel is equally likely to be fatal. The "blunt instrument" of legal parlance – which can be anything: a hammer or an axe, a baseball bat, a beam of wood, a rock, even a vacuum cleaner or a huge urn – will break bones, damage internal organs, fracture the skull and injure the brain, and, at the very least, leave its mark upon the flesh.

Bruises

It might seem that few physical injuries could be of greater significance to the forensic pathologist than bruises. Writers of detective stories – and even prosecuting counsel in court – have often suggested that a bruise can accurately reveal the point of injury, the force exerted, even the shape of the object causing the injury. The truth, however, is seldom as simple as that.

A bruise – known medically as a contusion – is an escape of blood into the tissues, due to the rupture of some of the smaller blood vessels, usually the minor veins or capillaries. The first point to remember about a bruise, therefore, is that it can be inflicted only on a living person, since blood will not flow from the capillaries after death. Violent attacks on a corpse can produce injuries resembling bruising, but these will be relatively small in comparison with the force employed. Examination at autopsy should soon reveal that the

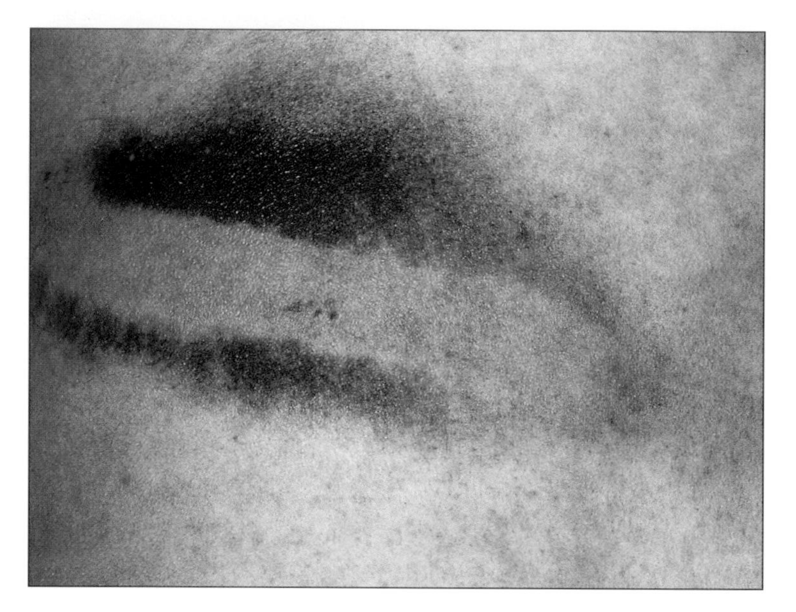

Bruising to the ribs. Because the bone lies close beneath the skin, the tramline bruise marks appear on either side.

injury is different from a true bruise.

The second important point is that, although it may precede death, the bruise is not in itself the cause of death. Bruises are of significance because they help to indicate the circumstances of death or severe injury. And – in certain instances – the object that produced the injury can be suggested by the nature of the bruise. Moreover, in cases of rape, or other forms of criminal assault, the position and form of bruises often provide some of the most telling evidence.

Blood vessels rupture because of intense local pressure exerted between the object causing the injury and the underlying bones, and so bruises must be carefully distinguished from abrasions or lacerations. The leaking of blood often occurs in quite a shallow area under the skin but, since the underlying bone resists the blow with an equal pressure, deep bruising can occur in any intervening tissue or organ.

The leaking blood usually diffuses through the tissues, following the fascial planes (the layers of tissue under the skin or between the muscles), and because of this it will most often not reveal the shape of what caused the bruise. An exception to this is "intradermal bruising"; this takes place only in the uppermost layer of tissue just under the skin, and it may therefore reproduce the pattern of the object that caused it. One can see this effect when the skin has been squeezed into grooves, such as by the tread of a car tyre, or when the victim has been hit with a patterned object such as a plaited whip or a decorative belt.

One intriguing case of patterned bruising occurred in Yorkshire, in the north of England. A miner was killed in an accident at the coal mine, and his trunk was covered by parallel zig-zag bruises. At first it was thought that these had been caused by the belt of the coal-conveyor that had crushed him to the ground, but the weave of the belt had a very different pattern. A technician at Leeds City Mortuary solved the problem; he pointed out that the bruises matched the inside pattern of the knitted sweater worn by the dead man.

Beating with a smooth thin rod often produces "tramline" bruising: two parallel lines caused by the sides of the rod, rather than by the direct blow. The blood vessels under the point of impact are squeezed by the blow and emptied of blood, while those on either side rupture.

Most bruises, however, are round or oval in shape, as the blood leaks fairly equally in all directions from the point of injury, and can be anything from

CRIME FILE:

Neville Heath

Good-looking and persuasive, he introduced himself as a high-ranking military man. It was the distinctive pattern of the riding whip with which he had brutally beaten one victim that became the final evidence against this murdering sadist.

Neville Heath, in police custody, following his arrest for the murder of Margery Gardner.

The asphyxiated body of Margery Gardner was found in bed at the Pembridge Court Hotel, in west London, England, on the afternoon of June 20, 1946. The examining pathologist, Dr. Keith Simpson, quickly established that she had suffocated. What interested and appalled him, however, were the injuries she had sustained before her death. Both breasts had been savagely bitten, there was a 7-inch (17.5-centimetre) tear in the vagina, and there were 17 clearly defined lash bruises made by a leather riding whip, plaited in a distinctive diamond pattern. "If you could find that whip you've found your man," Simpson told the police.

The room had been taken on June 16 by a man registering himself as Lt-Col. Neville Heath, but he was long gone. Some days later, giving the false name of Group Captain Rupert Brooke, he booked into a hotel in Bournemouth. There, on July 3, he had dinner with Doreen Marshall, who was reported missing two days later. Strangely, "Brooke" then visited the local police station and confirmed that he had dined with her. But his photograph had been circulated, and he was soon recognized and detained. Heath asked for his jacket to be fetched from his hotel, and in the pocket the police found a left-luggage receipt for his suitcase at Bournemouth Station. It contained a bloodstained scarf with which he had suffocated Margery Gardner, and a leather whip plaited in a diamond pattern.

A few hours later, the body of Doreen Marshall was found in a clump of rhododendron bushes. She was naked, her throat had been cut, and her body was savagely mutilated. At Heath's trial for murder, the jury took just an hour to find him guilty, and he was sentenced to be hanged. Among the evidence was the distinctive whip that Simpson had predicted would be his undoing.

Police examine the site at which Margery Gardner's body was found in Bournemouth, England.

CASE CLOSED

a fraction of an inch to several inches across. They are slightly raised above the surface of the skin by the accumulated blood, and this is the first characteristic to distinguish them from apparent bruises caused after death.

If the victim remains alive, even for only a short time after receiving the blow, the blood will continue to escape into the tissues, and therefore the size of the bruise is likely to be larger than the area of the surface that caused it. If the injury caused enough blood to be released on impact, this blood will continue to diffuse through the tissues after death. Depending upon the position of the body, it can move considerable distances, both in the living and the dead. It can diffuse towards the surface of the skin – the familiar effect of a bruise "coming out" – or through the lower tissues: a bruise on the thigh may subsequently appear at the knee, and an injury high on the scalp can reveal itself as a black eye.

As time passes, a bruise will change colour, due to breakdown of the hemoglobin in the blood: from red it rapidly goes to blue-black, and then to brown, green, yellow, and finally fades. It is not possible to date a bruise with any accuracy, since even two bruises in the same individual may change colour at different rates. In general, a bruise will take a week or two to go through its spectrum of colour changes, although some may fade, in a fit person, within three or four days. Nevertheless, it is important, particularly in cases of alleged child abuse, to note when bruises of different colours occur on one body. Those in charge of a child usually claim that all the bruises are the result of a single accident, but if some bruises are of a brown to yellow colour they cannot have been sustained within the previous 24 hours.

On occasions, bruising can appear far from the point of injury. A heavy blow to the scalp, for instance, can reveal itself as a black eye.

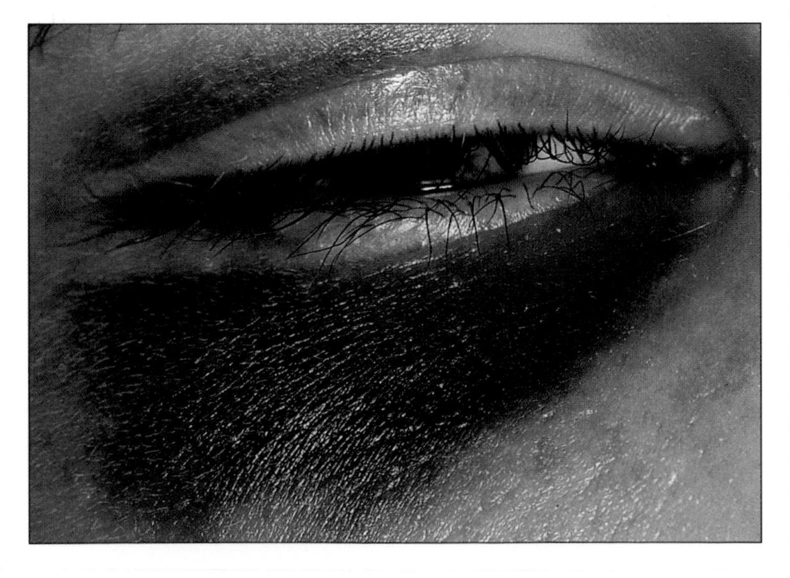

Bruises show up most clearly over prominent parts of the body, but the pathologist has to give particular attention to other areas. In cases of strangulation, surface bruises caused by fingers in the neck may be faint and small; on the other hand, they may be much larger than the pads of the fingers of the assailant. Bruises, in such cases, must also be searched for deep in the neck muscles, as they may not be externally visible.

When the shoulder blades show bruising, this is evidence that the body was pressed against the ground or some other surface, such as when the assailant knelt on his victim while throttling him or her. Bruising of the arms shows that the victim was being forcibly restrained. Rape victims usually sustain bruising of the inner thighs, and sometimes of the vulva, together with bruising of the face and arms as evidence of a struggle.

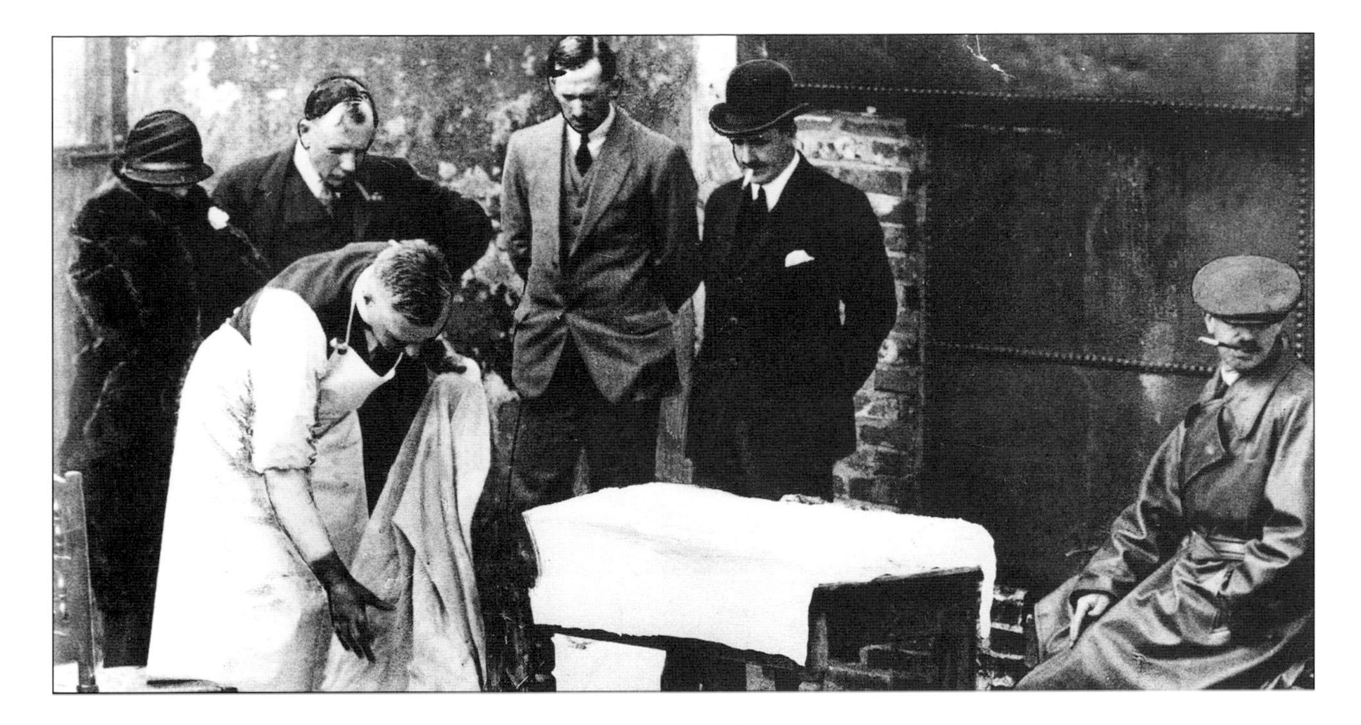

Bernard Spilsbury at work, examining the remains of Emily Kaye, murdered and dismembered by Patrick Mahon in a bungalow on the outskirts of Eastbourne, Sussex, in April 1924.

It is easy for the pathologist to overlook bruises of the scalp, which are usually covered with hair. These bruises can sometimes be felt with the fingers, even if they are not visible. If there is any suspicion of bruising to the head, the scalp must be carefully shaved, and the whole area of the skull examined minutely.

Although it is a useful rule of thumb that a heavy blow produces a larger bruise than a light one, it is in practice difficult to determine the violence of the blow from the appearance of the bruising. Large superficial bruises of tissues such as the eyelids and the external genitals can be caused by mild force, whereas tissues close to the bone, as on the scalp, bruise only with a blow of considerable strength. The very young and the aged, the fat, and those in poor physical condition all bruise more readily. A gentle squeeze, given in play to a plump, healthy, female arm, can produce a bruise that looks like the result of forcible restraint.

On the other hand, even considerable violence may leave no trace of bruising. The eminent English pathologist Sir Bernard Spilsbury wrote that this was true of up to 50 percent of severe abdominal internal injuries, where the blow caused rupture of a vital organ without rupturing the blood vessels at the point of impact. Also, if pressure is maintained until after the death of the victim – when, for example, an assailant keeps his stranglehold on the neck of the victim, or the wheel of a car remains in contact with the body – visible bruising may not develop.

Finally, there is one type of bruise in which the rupture of the blood vessels is caused, not by pressure, but by suction. This is the "love bite", which may be found on the neck, breasts or other parts of the body in cases of sexual assault.

CRIME FILE:

Karen and Michael Diehl

Travelling through America with 17 children in a converted bus, the couple had a hard time maintaining discipline, and frequently resorted to beatings. But one day they took a step too far, and a 13-year-old boy, Andrew, died from blows to the head.

Karen Diehl, accused with her husband Michael of the murder of their adopted son, Andrew, arrives at court in Virginia Beach on December 17, 1986.

Christian fundamentalists Karen and Michael Diehl were unusual parents. They travelled America in a converted school bus with their "family" of four natural and thirteen adopted children. They had decided deliberately to adopt children who were considered unadoptable.

In September 1986, the bus had arrived at a campsite in Virginia Beach, Virginia, at the mouth of Chesapeake Bay. On the morning of October 24, police and paramedics received an urgent call from the Diehls; when they arrived, they found one of the adopted children, 13-year-old Andrew, in a condition of cardiac arrest. Michael Diehl stated that the boy had been walking in the aisle of the bus when he collapsed and struck his head, either on the edge of a bunk bed, or on a box on the floor.

Andrew's heartbeat was stabilized, and he was rushed unconscious to hospital. There, doctors discovered that his bare buttocks and left eye were bruised, his lower lip cut, his ankles, feet, wrists, and hands blistered and swollen, and his torso and limbs covered with multiple old scars. A scan of his brain revealed a large subdural hematoma – a pool of blood between the skull and brain – on the upper right side of the head, and massive swelling of the brain itself. After five days in a coma, Andrew died.

At autopsy, the medical examiner recorded "a large red and yellow bruise on the top of the head, and smaller red bruises over the left eyebrow and on the back of the head". There were no cuts in the scalp, or fractures of the skull, but, on opening the skull, he found the large bulging pool of blood that had been detected in the scan. He also noted that the brain had softened and shifted to the left, and that there were contusions and dead tissue in several areas. The boy had a very low blood platelet count, and this was why the hemorrhage had refused to clot.

The medical examiner concluded that the cause of death was "head injuries due to blows". He stated that "the location of the head injuries is inconsistent with a typical fall"; and added that the abrasions on the boy's wrists were "consistent with binding", that the contusions on his buttocks were "consistent with the whip recovered at the scene", and that a scar on his chest was "consistent with blows from a linear object". Following this, Karen and Michael Diehl were each charged with child neglect, assault, abduction, and murder.

Inquiries revealed that the Diehls had regularly beaten the children, who were admittedly unruly, with a thin two-foot wooden paddle. Andrew had been particularly undisciplined, deliberately wetting and soiling his bed and other children's clothes. Eventually, he was made to sleep naked on the rubber matting of the bus with his wrists handcuffed.

On the evening of October 22, Karen Diehl said, she had "tapped" Andrew on the top of his head with the paddle as he sat on the toilet. When she bathed him on the morning of October 24, she noticed that his feet were swollen; he had difficulty walking, and she assisted him up and down the aisle of the bus. She had only turned her back for a moment to speak to her husband when Andrew fell and cut his lip. After wiping it, she left him where he was, because "he appeared to be lying there comfortably". A few minutes later, he was unconscious and his breathing was irregular, but it was 45 minutes before the Diehls summoned medical assistance.

Dr. Cyril Wecht, one of America's leading forensic experts, was retained by Karen Diehl's defence. He inspected the bus, and noted that, above the toilet, the roof curved so that only a foot or so of space would have been above Andrew's head as he sat there. "I tried to picture how anyone could swing a paddle in that manner with enough force to cause the kind of internal bleeding Andrew had suffered. It just did not seem possible."

In the court hearing, there was little disagreement about the strict disciplinary methods practised by the Diehls, and the defence concentrated on whether the blow with the paddle, which was also undisputed, could have resulted in the hemorrhageing that appeared to be the cause of death. The neurosurgeon from the hospital where Andrew had died gave his opinion that the hemorrhage was due to "blunt injury to the head", basing his conclusion on the absence of skull fractures or scalp lacerations. Cross-examined, he agreed that the injury to Andrew's left eye was consistent with a fall, and admitted that the hemorrhage could have been caused by a combination of a fall against a bunk and the low platelet count.

Questioned by the prosecutor, the medical examiner explained why he did not think the hemorrhage had been caused by a fall. Fall injuries usually lie in a "hat-band" distribution, whereas one of the bruises was on the top of the head. Moreover, this bruise extended over the curved surface of the scalp, which he felt corresponded with "more than one impact, which is extremely atypical for a fall". Explaining the absence of surface injuries to the scalp, he said that "the hair acts in some ways as a lubricant", allowing a blow "to essentially slide off without producing near as much abrasion as the same injury might produce in other locations".

The prosecutor then asked the medical examiner to take the paddle and demonstrate the degree of force that had been used, to which he replied: "I think I'm going to need something to hit. I don't want to damage the furniture". He added that probably more than one, and possibly as many as three, blows had been delivered.

When Dr. Wecht was called, he testified his opinion that the subdural hematoma was the result of two or more falls. When defence counsel repeated the medical demonstration of the blow with the paddle, he replied: "Those kinds of blows did not produce the trauma in this child".

He repeated his opinion that the injuries "were caused by the two or more falls with injuries to the head, and were not consistent with forceful, direct blows against the top of the head in the fashion that has been demonstrated in this courtroom".

Despite Dr. Wecht's evidence, however, Karen Diehl was found guilty of involuntary manslaughter – while Michael Diehl was convicted of first-degree murder.

CASE CLOSED

At autopsy, bruises must be searched for in any area where it is suspected they might occur. Incision of the area will reveal the characteristic discoloration and tissue damage, and it is not difficult for the experienced pathologist to distinguish this from post-mortem hypostasis (see "Gathering the Evidence") or injuries caused by the dead body being struck or dropped. There is also likely to be a concentration of white blood cells (leucocytes) in the area. A tissue section is therefore always taken for microscopic examination. Some experts claim that another type of blood cells, the phagocytes, begins to form a decomposition product of hemoglobin called haemosiderin after 24 hours, and that this can be detected. Equally, it may be present as the result of a previous injury.

Bruises, therefore, cannot be relied upon in evidence as an infallible indication of the cause of death, or even, necessarily, of the infliction of violence. They can, however, provide valuable circumstantial evidence, whenever it is possible to establish what has caused them.

Lacerations

When the skin is split by the strength of a blow, the wound is known as a laceration. The tissues at the edge are grazed and bruised, and the split is irregular, with strands of nerve tissue and blood vessels stretching across it. Death from blows with a blunt instrument is generally due to fracture of the skull. A weapon with a point, such as a hammer, an axe or the corner of a brick, will cause a depressed fracture, driving fragments of bone inward. Even if the skull is not broken, the brain can be badly damaged, and there may be fatal hemorrhaging.

A single blow from a metal rod can produce a laceration that is Y-shaped. Oddly, a single blow can also result in more than one laceration. A blow on the side of the head, for example, can result in laceration of the lower jaw, the ear, and the front of the brow. On the other hand, blows to lower parts of the body may not lacerate the skin, but cause extensive laceration of the soft tissue beneath.

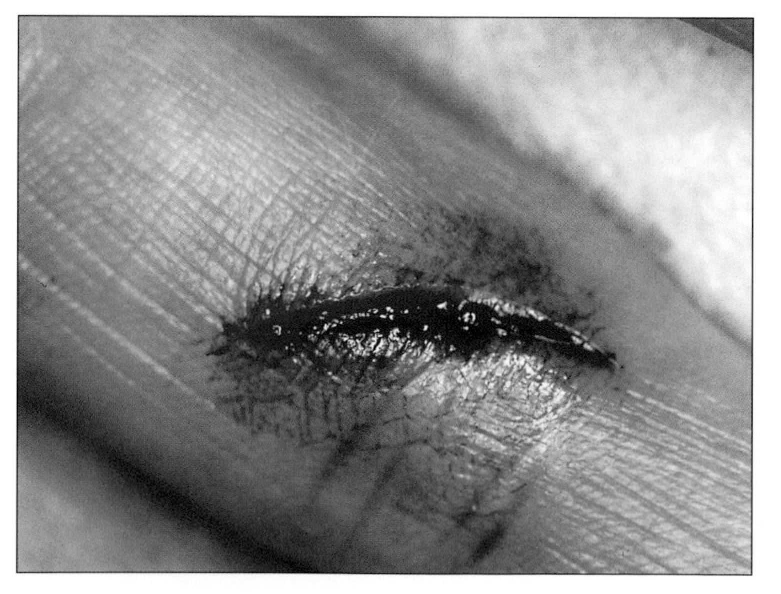

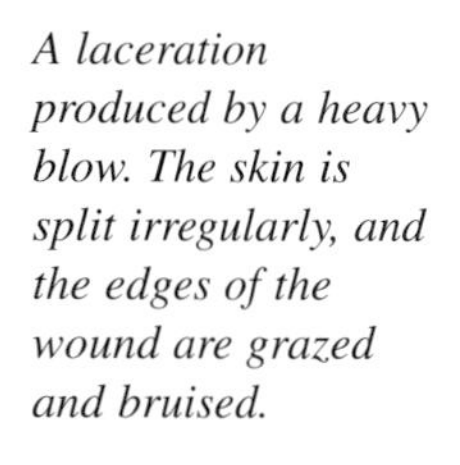

A laceration produced by a heavy blow. The skin is split irregularly, and the edges of the wound are grazed and bruised.

Knife Wounds

When a body is found, killed with a knife or other sharp implement, the visible evidence is there for all to see: blood, the wound or wounds – and even the weapon is often left at the scene of the crime. There are cases in which death has been caused by some other means – strangulation, smothering or the use of a blunt instrument, and the murderer has subsequently slashed or stabbed the body – but these will be characterized by relatively little loss of blood from the wounds.

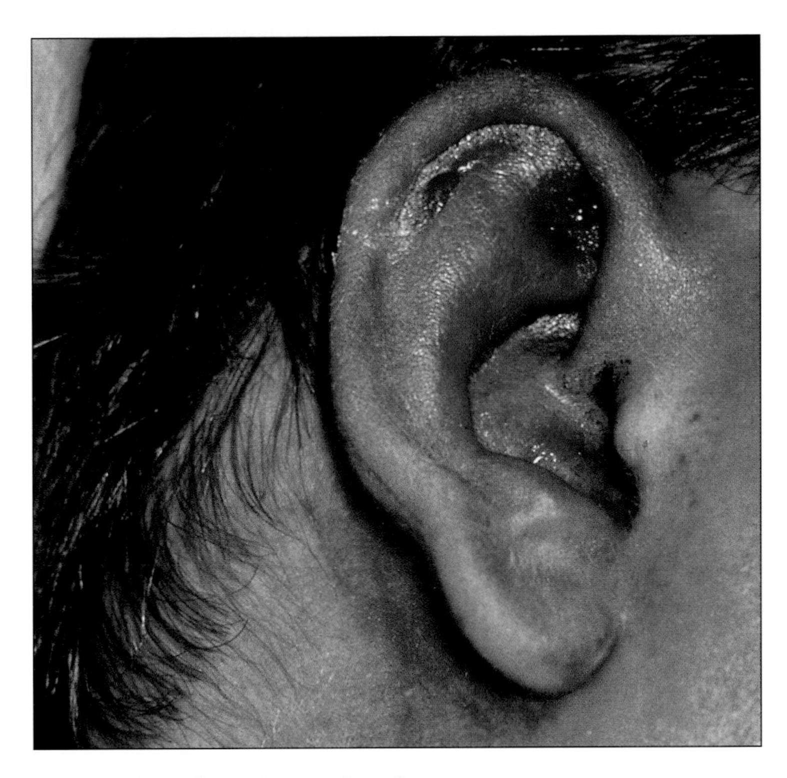

A blow to the scalp will often result in blood migrating to another part of the head. Here the hemorrhaging has become visible in the ear.

Wounds caused by a sharp weapon can be of two types: incised wounds, produced by slashing with a razor, the blade of a knife, a jagged piece of metal or a piece of broken glass; and punctured wounds, produced by the point of a knife or other long, narrow, instrument.

Incised wounds are usually straight, but may be curved or V-shaped if the direction of the weapon is changed, or if the blade of the weapon has an unusual shape: a curved pruning knife, for example. Wounds caused by jagged metal or glass may appear at first to be irregular, but closer examination reveals that the edges have been cleanly cut.

Incised wounds usually gape, and so the width of the wound cannot be matched with the breadth of the cutting edge that produced it. In a deep wound, muscles, tendons, nerves and blood vessels may all be severed, and the cutting of the muscles causes the wound to gape even wider.

Often the victim tries to ward off the blows of the knife, or attempts to seize it, and there are defence cuts on the forearms and the palms of the hands.

As with incised wounds, it is a fallacy to believe that the external appearance of a stab wound necessarily corresponds to the shape and dimensions of the weapon. As the great German criminologist Hans Gross wrote in his historic work *Criminal Investigation*:

> When the point of a knife penetrates into the body to a depth of half an inch or less, it forms at first a wound with a sharp or pointed angle at each end; as the knife proceeds further in, the end in contact with the cutting side of the knife naturally remains sharp and pointed; but the other end which is in contact with the back of the knife remains so also.

If the knife is twisted as it is pulled out after stabbing, the wound may have a V- or cross-shaped appearance. The size of the opening may also be smaller than the dimensions of the weapon, because the skin may be stretched by the pres-

CRIME FILE:
August Sangret

Joan Wolfe's skull had been smashed into nearly 40 pieces with a heavy wooden stake, but she had also been brutally attacked with a knife before she died. Finding the knife, and matching it to the wounds, was an essential part of the case.

When the body of Joan Wolfe was found on a common near Godalming in Surrey, England, on October 7, 1942, Dr. Keith Simpson was summoned to examine it. Her skull had been fractured into nearly 40 pieces, but when it was reconstructed three stab wounds were found in the front of her head.

The first question was whether these wounds had been made before or after death. There was another stab wound in Wolfe's right forearm below the elbow, and one in the palm of her hand: examination of the tissue showed that she had been alive when these were inflicted. It was therefore reasonable to infer that the head wounds had also been inflicted during life. They were close together, high on the left-hand side of the skull, indicating that they had been made by a right-handed person striking downward.

Joan Wolfe was killed by a heavy blow with a stake to the back of her head, and her skull fractured into nearly 40 pieces, but there were also three knife wounds to the front of her head.

A fragment of muscle had been pulled out of the wound in Wolfe's forearm, and a tendon had been similarly hooked from the palm of her hand. Dr. Simpson deduced that "the point of the weapon must have been something like a parrot's beak. The three holes in the skull

A sad memorial to a doomed relationship. Joan Wolfe wrote these words on the wall of a hut on a nearby cricket ground, where she and Sangret used to meet. It reads: "A. Sangret, Canada. J. Wolfe now Mrs. Sangret. England, Sept. 1942 (left); "My love lies over the ocean. Please bring him back to me." (right).

When Joan Wolfe's half-buried body was discovered on October 7, 1942, parts had already mummified, and adipocere had formed in her breasts and thighs.

vault were rimmed or bevelled .... it looked as if the beak-like point of the weapon had been driven into the head and turned or twisted before withdrawal."

Suspicion fell on Wolfe's boyfriend, a Native American named August Sangret, who was stationed nearby with the Canadian Army. However, no knife was found among his belongings, and the type of bone-handled jack-knife issued to the Canadians did not match Dr. Simpson's description of the weapon.

Witnesses connected Sangret with a British army issue knife with a hooked tip, but it could not be found. A month passed before it was eventually discovered in a drain at the Canadian camp. At Sangret's trial, Dr. Simpson demonstrated how it exactly fitted the holes in Joan Wolfe's skull.

CASE CLOSED

Medical witnesses in court are often asked about the force needed to produce a particular stabbing wound. A number of factors have to be taken into consideration.

- Most important is the sharpness of the point of the weapon. The skin is the body's most resistant tissue and, once it has been penetrated, the sharpness of the rest of the blade is less important.
- The more rapid the stabbing movement, the easier it is to penetrate the skin.
- Once the point of the weapon has penetrated the skin, almost no additional force is required to penetrate the tissues.
- Where the skin is stretched tightly across the ribs, pressure from the little finger alone is sufficient for penetration with a sharp-pointed weapon, and the heart and other organs are far less resistant to penetration.
- A person can easily stab him- or herself by falling or walking on to a sharp weapon held by another person; no movement of the knife is necessary, and it does not even have to be held rigidly.

sure of the point before penetration. On the other hand, it can also be larger if the weapon is pulled out obliquely.

A rounded sharp weapon may not produce a circular external wound, because the skin may split in one direction; a file or a bayonet, or a square-sectioned spike, can produce a triangular or cross-shaped external wound. A closed pair of scissors may leave a "stepped" wound shaped like a lightning flash.

For these reasons, it is essential for the pathologist to examine the interior of the wound, and the damage sustained by underlying tissues, before making any estimate of the shape and dimensions of the weapon. Even the depth of the wound can be misleading: it may be greater than the length of the weapon, due to the tissues being compressed during penetration.

A stabbing weapon can be held in two ways. A knife held point downward, with the thumb round the upper part of the hilt, is less likely to produce a lethal wound, particularly if assailant and victim are facing one another. This is because the attacker uses a downward thrust, and the point of the knife is liable to strike bone before reaching a vital organ such as the heart or lungs. Far more dangerous is a knife held with the point upward, with the thumb close to the blade, as in a flick-knife. Careful observation of the direction of the wound is therefore of great importance in determining the relative positions of the assailant and victim.

There may be only a limited amount of external bleeding from a puncture wound, but serious internal bleeding in a wound in the chest or abdomen. Faced with a dead body with incised or punctured wounds, the pathologist still has to ascertain whether it is a case of homicide or suicide. Suicide is most likely to be attempted by cutting the throat or the wrist, and there are certain signs to look out for.

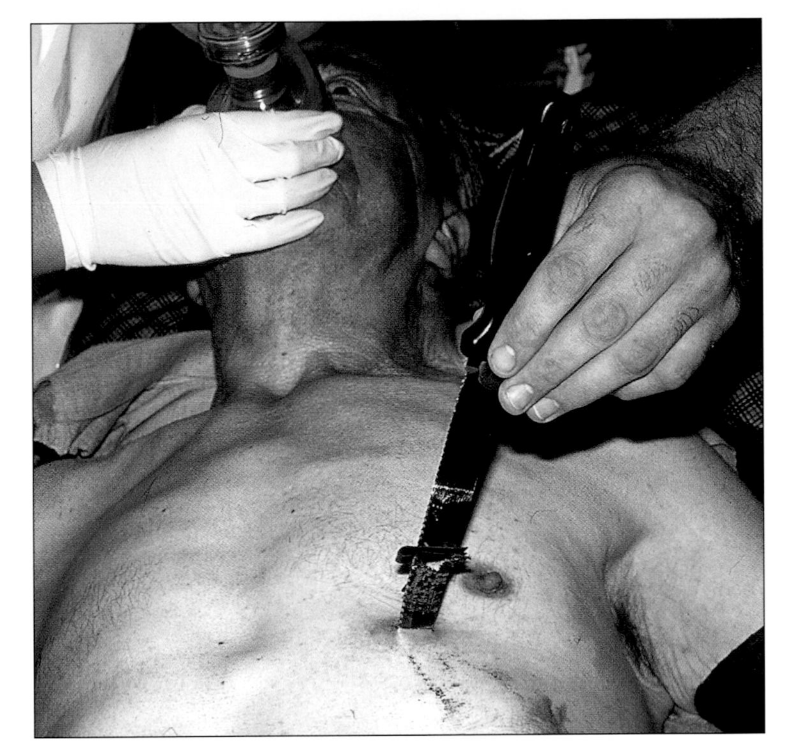

When a right-handed person cuts his or her throat, the wound normally begins high up on the left, where it is deep, and finishes lower on the right, where it tails off; a wound like this could be inflicted by another person only by standing behind the victim, and would probably be as deep, if not deeper, at the right. The homicidal wound is also lower in the neck, and probably more horizontal.

The self-inflicted wound is likely to be clean-cut, because the suicide tends to throw back his or her head, stretching the skin of the throat, before slashing. A murder victim taken unawares is usually relaxed, and the skin folds under the pressure of the blade, giving an uneven margin to the wound.

Also, in a suicide one generally finds several separate, shallow cuts near the upper end of the wound, where the person has made tentative, hesitant attempts before finally slashing the throat. These are absent in a homicidal wound, which is likely to be accompanied by other deep cuts on the head or neck. Tentative wounds are particularly characteristic of a suicide performed by cutting the wrist, and are generally parallel to one another. Defence wounds, produced by attempts to grab or ward off the weapon, are, however, random, usually on the palms of the hands or the knuckles, and often on the forearms.

In cases of stabbing, whether intentional or accidental, the knife should never be extracted from the wound, except under strict medical supervision. Careless removal can be followed by massive bleeding, and could result in death.

Very occasionally, a suicide may cut his or her neck from behind. In one case a butcher, who had failed to kill himself by slashing his throat, used the method, at the back of the neck, with which he was familiar from slaughtering animals.

Stab wounds can also be suicidal, although they are more likely to be homicidal. Here the track of the wound in the tissues must be determined carefully. It is easier to stab oneself in a downward or horizontal direction; wounds of this kind may occur accidentally, by running or falling on to the weapon. Upward thrusts generally indicate a homicidal attack. The direction of the wound is also important, if one knows whether the dead person was left- or right-handed.

The site of the wound is also significant since, if it is in a position that the victim could not reach, it could not be self-inflicted. In both suicide and homicide, the heart is generally the target, although homicidal wounds tend to be higher than suicidal ones. Finally, a suicide is usually successful first time – setting shallow tentative wounds aside – and multiple deep wounds are a clear suggestion of murder.

With Poison Deadly

IT IS A FACT – NOT WIDELY UNDERSTOOD – that any substance, whatever its nature, can act as a poison. Even water, if consumed in sufficient quantity, can kill. The fact was recognized as early as the 16th century by the remarkable physician and alchemist Theophrastus Bombastus von Hohenheim, who called himself Paracelsus (meaning "without equal"). He wrote: "All substances are poisons; there is none that is not a poison. The right dose differentiates a poison and a remedy." Death may be slow and lingering, or swift and sudden. Substances that rapidly cause death when administered in relatively small quantities, or that become lethal when cumulatively taken over a period, are commonly regarded as poisons, and may be used as instruments of homicide.

When one considers the wide range of poisonous substances available in the modern family medicine cupboard, it seems surprising that murder by poison is so uncommon. A significant number of cases may well go undetected if the victim's physician certifies that death is due to natural causes. This is particularly possible in circumstances where the victim is suffering from a chronic, possibly life-threatening, disease – there are many cases, for instance, of relatives who have admitted giving a drug overdose to a seriously ill patient, to end the suffering.

The glossy purple-black berries of deadly nightshade, Atropa belladonna. The poisonous alkaloids atropine, hyoscyamine and hyoscine can be extracted from this plant.

Moreover, even if the examining physician or pathologist suspects poisoning, a full toxicological analysis can be a long and laborious business, unless circumstantial evidence suggests a specific substance. As the distinguished English pathologist Professor Keith Simpson put it: "Homicide by poison is

CRIME FILE:

Marie Lafarge

Everybody was sure that she had poisoned her bankrupt husband with arsenic, but it took all the skill of Mathieu Orfila, the founder of toxicology, to make forensic history, and establish the first scientific proof.

Marie Lafarge. The murder of her husband in 1840 was the first case in which arsenic poisoning was scientifically demonstrated.

In August 1839 Marie Capelle was married, at the age of 23 – much against her will – to a middle-aged, bankrupt ironmaster named Charles Lafarge. Marie Capelle had dreamt of marrying a rich man with a place in society, not the owner of a dismal, rat-infested forge at Le Glandier, in the Limousin, France, and she was very unhappy.

In December, she bought some arsenic to kill the rats. Some days later, she sent a cake to her husband, who was away on business in Paris. Lafarge became violently ill, and when he returned to Le Glandier he fell sick again. His wife fed him with her own hands and was seen by a servant to stir a white powder into his food. Suspicions were aroused. Lafarge's family asked a local pharmacist to test the food, and he reported that it contained arsenic; Charles Lafarge died on January 14, and his wife was arrested.

When the trial opened in Tulle on September 3, 1840, experts retained by the prosecution announced that the Marsh test had not revealed arsenic in Lafarge's stomach. They asked for an exhumation so that other organs of the victim's body could be analyzed, but once again the tests proved negative. However, various foodstuffs from the Lafarge household were found to contain arsenic, including "enough to poison ten persons" in some eggnog.

Mathieu Orfila was asked to resolve the deadlock. He questioned all the experts closely, and examined the materials they had used in their tests. Then, behind the locked doors of a room at the Tulle courthouse, he performed the Marsh test correctly, and showed that the experts had bungled their findings.

Orfila's evidence settled the matter. "I shall prove," he said, "first, that there is arsenic in the body of Lafarge; second, that this arsenic comes neither from the reagents with which we worked nor from the earth surrounding the coffin; also, that the arsenic we found is not the arsenic component that is naturally found in every human body."

Marie Lafarge was found guilty and sentenced to death, but the sentence was later commuted to one of imprisonment with hard labour.

CASE CLOSED

rare. The Maybricks, Seddons, Crippens and Merrifields are famous only because they are of rare interest."

Arsenic and the Birth of Toxicology

Until the 19th century, there was very little scientific investigation of unnatural death. If the cause of death was obvious, and circumstantial evidence identified the perpetrator, that was generally sufficient for a verdict of murder. A case of poisoning, however, was a rather different matter.

For many centuries, poison remained a virtually undetectable way of killing someone – in the early 17th century, Phineas Foote described it as "the cowards' weapon". The agonies suffered by the victims, and their untimely death, frequently led to the suspicion that they had been poisoned, and circumstances often suggested the identity of the murderer, but proof was almost impossible to establish. In Imperial Rome, poisoning became so common, as a way of removing someone who was an obstacle to advancement, that many members of the upper classes employed food-tasters; in the Middle Ages the Italian Borgia family was said to have poisoned many of its enemies.

Ancient peoples were familiar with many different poisons, mainly from vegetable sources, and used them for practical purposes such as hunting. When Sophocles, for example, was found guilty of corrupting Athenian youth, he was given the opportunity to drink an extract of hemlock to bring about his death. In more recent centuries, an easily obtainable mineral, arsenious oxide – known commonly as "arsenic" – became the poison of choice for murder. Its faintly sweet taste was readily disguised when it was added to food, and its lethal effects were frequently attributed to acute gastric disease. There was no way to detect it in a dead body until the early 19th century.

The Minorcan Mathieu Orfila, founder of the science of toxicology.

The man who first established the study of poisons – toxicology – on a scientific basis was Mathieu Orfila. Born in Minorca in 1787, he won a scholarship to Spain's Barcelona University, and then studied in Paris, France, to gain his medical degree. There, when he attempted to demonstrate the accepted tests for various poisons, he discovered that they were entirely unreliable. As he later wrote: "The central fact that had struck me had never been perceived by anyone else … toxicology does not yet exist."

Orfila published his first volume on the subject, a *Treatise of General Toxicology*, in 1813. He quickly became famous, and in 1819 was appointed professor of Medical Jurisprudence at Paris University, where he was called upon to give evidence in a number of poisoning cases. In the course of his work, he wondered whether the soil of cemeteries might contain arsenic, which could find its way into buried bodies, and be found after a subsequent exhumation – so bringing the evidence of a toxicologist into doubt. He would have been unable to carry out an investigation of this question, had it not been for the delicate test developed in 1836 by the English chemist James Marsh (see Fact File below). Arsenic was revealed in some soils, but Orfila showed that it would not enter a sealed coffin. His experiments also proved that the various chemicals used in tests for arsenic might themselves be contaminated with arsenic, and that care must be taken to allow for this.

Arsenic was so freely available as a rat poison during the 19th century that it became the favourite means of murder, particularly among poorer people. Following the Lafarge case, the French authorities introduced a law forbidding pharmacists to sell arsenic to anybody unknown to them, and making it necessary for purchasers to sign a "poisons book". It was soon followed in Britain by the pass-

FACT FILE

It took the work of several chemists, over a period of 60 years, to develop a reliable method of detecting tiny traces of arsenic. The first of these was the great Swedish scientist Carl Wilhelm Scheele. In 1775 he dissolved arsenious oxide in nitric acid, and added zinc granules to it. The solution gave off a poisonous gas – later named arsine – that smelled of garlic.

Some years later, the German Johann Metzger discovered that when arsenious oxide was heated with charcoal, a mirror-like deposit would form on a cold plate held over it. This was in fact the element arsenic. Then, in 1806, Dr. Valentine Rose in Berlin, Germany, took the stomach of a suspected poison victim, treated it with nitric acid, added potassium carbonate and lime to the liquid, and evaporated it to a white powder. When he heated this with charcoal, he obtained the characteristic arsenic mirror. Rose's method was used in 1810 to prove that a domestic servant named Anna Zwanziger had poisoned several people who employed her.

The final development came a generation later. In 1832, an elderly farmer in England named George Bodle was poisoned by his grandson John. James Marsh, a former assistant of the famous scientist Michael Faraday, was asked to demonstrate that a sample of the man's coffee contained arsenic. He did so, but was unable to explain his work to the jury, who found John Bodle not guilty.

Marsh determined to find a way of presenting visible evidence in any future case. He realized that Metzger's technique was not sufficiently delicate, and went back to Scheele's discovery of arsine. The result was simple and elegant. He treated the suspect material with sulphuric acid and zinc in a closed bottle, and arsine evolved passed through a narrow glass tube that was heated over a short distance. An arsenic mirror formed further along the tube, and any escaping gas was burnt at a nozzle at the end, where it formed another mirror on a porcelain plate. As little as 0.02 milligrams of arsenic could be detected in this way. For this discovery, Marsh was awarded the gold medal of the Society of Arts in 1836.

The Marsh test is still taught to all chemistry students, but in forensic work it has been superseded by equally delicate tests developed by other chemists.

CRIME FILE:

Marie Hilley

Long after the "golden age" of arsenic poisoning, she murdered her husband and mother, and attempted the death of her daughter, with arsenic. Then, jumping bail, she faked her own death.

In 1975, in Anniston, Alabama, Marie Hilley's husband Frank died after a short illness, and the cause of death was given as infectious hepatitis. His wife claimed his life insurance – but it was soon spent. In July 1978, she insured the life of her 18-year-old daughter, Carol, for $25,000.

Not long after, Carol Hilley was confined to hospital with a mysterious disease, and came near to death. Her mother visited her regularly, bringing her special foods; several months later the daughter casually told a friend that her mother had also been giving her injections without the knowledge or consent of the doctors.

When Hilley was arrested in September 1979, on a charge of passing worthless cheques, she was also asked about her daughter's sickness. A sample of the daughter's urine contained arsenic; Frank Hilley's body was exhumed, as well as that of Marie Hilley's mother: both revealed a high arsenic content. Hilley was charged with her husband's murder, but jumped bail in November 1979, and was not apprehended until January 1983.

In the meantime, Marie Hilley – who had married again – faked her own death and allegedly given her body to research, emerged describing herself as her twin sister, Teri Martin. She was found guilty of the murder of Frank Hilley and the attempted murder of her daughter, and was sentenced to life imprisonment, plus 20 years.

CASE CLOSED

Marie Hilley, found guilty of the murder of her husband Frank in 1975, and the attempted murder of her daughter in 1978.

ing of the Arsenic Act of 1851. This restricted sale of rat poison and other arsenic products to persons over 21, who had to be known to the seller, and were required to sign the poisons register. Most importantly, the poison had to be mixed with either soot or indigo, so that the white powder could not be confused with foods such as sugar or flour. Other countries enacted similar legislation.

Nevertheless, arsenic continued to be used in murder. Helene Jegado was

Madeleine Smith in the dock. The charge of murdering her lover Emile l'Angelier with arsenic in 1857 was found Not Proven in the Glasgow court. She died in America in April 1928, and was buried in Mount Hope cemetery, New Jersey.

found guilty of three murders with arsenic – with 20 other likely cases – in France in 1851; Madeleine Smith poisoned her lover in Glasgow in 1857; Mary Anne Cotton killed 14 or 15 persons in England over a 20-year period up to 1873; Cordelia Botkin murdered her lover's wife and sister-in-law in San Francisco in 1898; Herbert Rowse Armstrong poisoned his wife, and attempted to murder a neighbour, in Hay-on-Wye, Wales, in 1921; Mary Creighton and her lover Everett Appelgate used arsenic to kill Appelgate's wife on Long Island, New York, in 1935.

Apart from its presence in the internal organs, arsenic can also be detected in the hair and fingernails, particularly in cases of chronic poisoning. Since both grow at a regular rate, detection can help determine the times and period during which the poison was administered. In a classic experiment – upon himself – Dr. Alan Curry, at that time head of the Home Office Forensic Laboratory at Aldermaston in England, discovered that it took 103 days for a dose of arsenic to grow out from the tip of his thumbnail.

Everett Appelgate (above) and his lover Mary Creighton (right centre) were found guilty of murdering Appelgate's wife with arsenic on Long Island in 1935.

The American explorer Charles Hall died on November 7, 1871 aboard the *Polaris*, in the course of a search for the North Pole. For nearly a century, there was doubt about the circumstances surrounding his death – had the expedition's scientist, Dr. Emil Bessels, poisoned him? In August 1968, Professor Chauncey Loomis, of Dartmouth College, New Hampshire, and pathologist Dr. Franklin Paddock flew to Hall's burial site, on the frozen shore of Thank God Harbor, about 500 miles from the Pole. They exhumed the body, which was remarkably well preserved, and took samples. A fingernail was subjected to neutron activation analysis (see "The Forensic Hardware") at the Toronto Center of Forensic Sciences. At the tip, its arsenic content was 24.6 parts per million (ppm), and at the base it was 76.7 ppm.

CRIME FILE:

Eva Rablen

Like a real-life Sherlock Holmes, Dr. Edward Heinrich, the "Wizard of Berkeley", deduced the only possible evidence that could link Eva Rablen directly to the poisoning of her war veteran husband, Carroll.

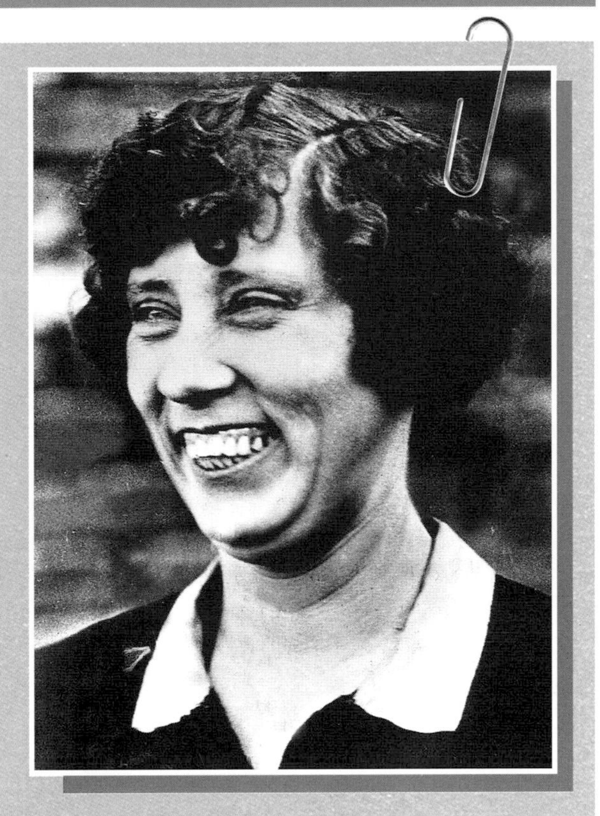

Eva Rablen, who poisoned her husband Carroll with strychnine in 1929. Her guilt was established by forensic scientist Dr. Edward Heinrich.

Carroll Rablen was a veteran of World War I, rendered deaf by a wound, but he found a modified pleasure in watching his wife Eva enjoying herself in dances at the town schoolhouse in Tuttletown, California. On April 29, 1929, he sat outside in his car, until his wife brought out a tray of coffee and sandwiches. A few minutes later, the dancers heard a scream, and discovered Carroll Rablen writhing in the agony of death.

Eva Rablen suggested that her husband had committed suicide; various other people – including the medical examiner – believed the death to be due to natural causes. Autopsy, and analysis of the victim's organs, revealed nothing.

Rablen's father, however, thought that Eva Rablen had poisoned her husband for his $3,000 life insurance. On his insistence, Sheriff Dampacher agreed to search the schoolhouse premises, and discovered a small bottle labelled "strychnine", hidden in a small dark space. The strychnine had been sold by a drugstore in Tuolumne; a short distance away, the clerk identified the purchaser as Eva Rablen, and she was arrested.

The prosecution called on the services of Dr. Edward Heinrich. As an independent forensic scientist, Heinrich had many remarkable feats of detection to his credit – he was known as the "Wizard of Berkeley". He analyzed Rablen's remains specifically for strychnine, and obtained a positive indication. He also found traces of the alkaloid in the cup from which he had drunk, and in the upholstery of the car.

There was still nothing to tie Eva Rablen directly to the poisoned coffee. Then a thought struck Heinrich: she had made her way across a crowded dance floor – might she not have bumped into somebody, and spilt a little from the tray? Sure enough, one young woman remembered such an incident, and still had her dress with coffee stains on it. And, sure enough, Heinrich discovered that those stains contained strychnine.

Heinrich's fame was so great that, when Rablen learned of his involvement in the case, she preferred to plead guilty, so avoiding the death penalty. Instead, she was sentenced to life imprisonment.

CASE CLOSED

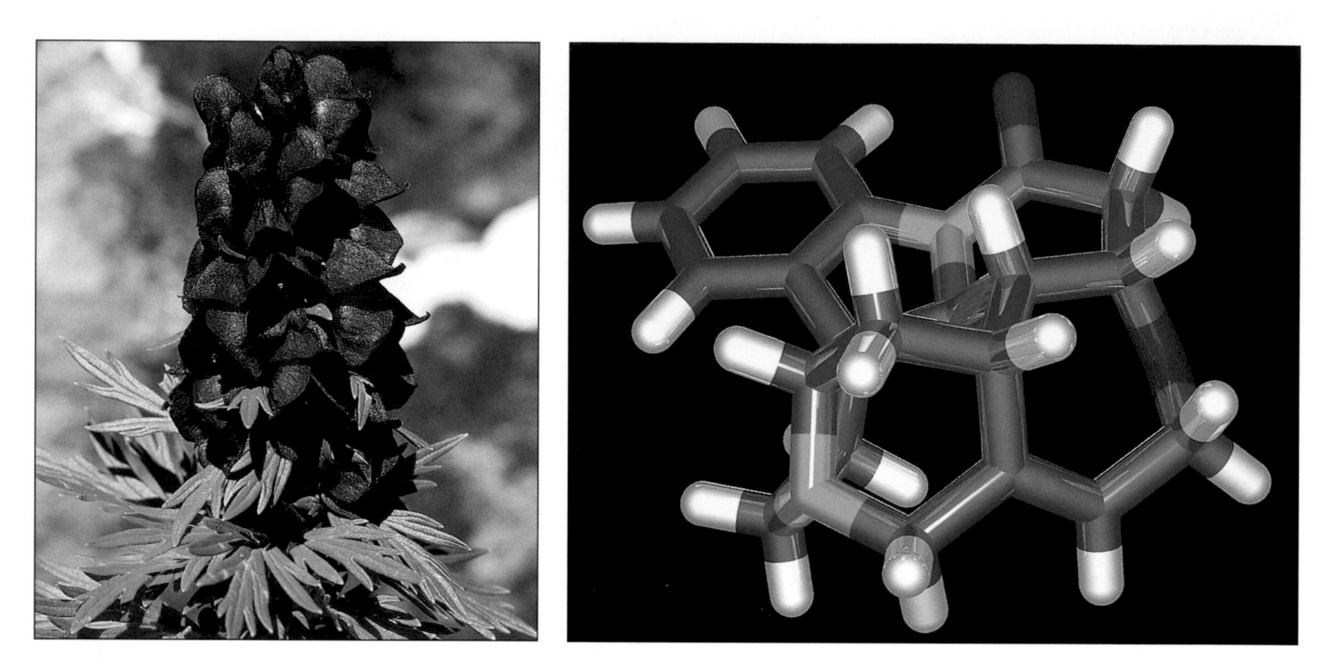

Above left: Wolfsbane, or monkshood, was a popular poison in Classical times. Above right: An idealised model of the molecular structure of strychnine. This alkaloid poison, extracted from the nux vomica, has no known antidote.

Although the soil surrounding the body contained 22 ppm, there seemed to be no reason why arsenic in the soil should not have been absorbed equally into all parts of the nail. The analyst declared that "arsenic poisoning is a fair diagnosis".

The "Golden Age" of Homicidal Poisoning

In the early years of the 19th century, chemistry developed from a largely hit-or-miss experimental procedure into a logically organized science. Scientists identified and isolated more and more elements. Also, compounds of all kinds were separated in a pure state from their natural raw materials. Precise methods of analysis were defined, and many chemists investigated the synthesis of naturally occurring substances. One outcome of this was that the stock-in-trade of pharmacists came to include an ever-increasing number of purified poisons.

The drug opium was widely used throughout the 19th century; its active ingredient, morphine, was isolated in 1803. Belladonna, the extract of deadly nightshade, was another common drug; atropine was isolated from it in 1833. Strychnine was obtained from the nux vomica tree in 1818, and aconitine from monkshood in 1833. Nicotine, a poison as deadly as hydrocyanic (prussic) acid, was isolated from tobacco

Queen Victoria's physician Dr. John Snow's special chloroform inhaler.

CRIME FILE:

Adelaide Bartlett

Did she poison her husband with chloroform? And, if so, how did she do it? Adelaide Bartlett kept her secret, the experts were stymied, and the jury had no alternative but to acquit her of the charge of murder.

After 11 years of apparently contented marriage, the 40-year-old Englishman Edwin Bartlett was found dead in bed on New Year's Day 1886. Subsequently, some odd facts emerged about the Bartletts. Edwin's French-born wife Adelaide was just 19 when she married him. Within a year, she had had an affair with his brother. Furthermore, according to her account, her husband showed little sexual interest in her, and even encouraged her friendship with a young Wesleyan minister, Rev. George Dyson.

The Bartletts moved into an apartment in London's Pimlico in October 1885, and Dyson visited them there regularly. Edwin Bartlett made a will in which he left everything to his wife, with Dyson as executor, and soon after fell ill. At his death, an autopsy revealed a large quantity of chloroform in his stomach, although no traces could be discovered in his mouth or throat.

The trial of Adelaide Bartlett for murder opened on April 12, 1886. Evidence in court revealed that Dyson had bought quantities of chloroform from various chemists in Putney and Wimbledon only days before Bartlett's death. However, his wife said that she had used it, sprinkled on a handkerchief, as a way of making her husband (who had suddenly shown an awakened "interest" in her) sleepy at night.

The cross-examination of Rev. George Dyson (inset) during the trial of Adelaide Bartlett (above), who was accused of murdering her husband Edwin with chloroform.

The problem facing the prosecution was explaining how the poison had been administered. Liquid chloroform will blister the mucous membranes if taken by mouth; it is soluble in water at 1 part in 200. To introduce the quantity found in the victim's stomach his wife would have had to persuade him to drink several pints of the mixture. Did she contrive to render him unconscious, with his compliance, and then introduce a rubber tube into his stomach?

In the light of insufficient evidence, the jury acquitted Bartlett. After the trial the surgeon Sir James Paget allegedly remarked: "Now she's acquitted, she should tell us, in the interests of science, how she did it". Adelaide Bartlett, however, kept silent.

CASE CLOSED

CRIME FILE:

Arthur Ford

Foolishly infatuated, he tried to attract a fellow employee by giving her the alleged aphrodisiac known as "Spanish fly". But his plans went badly awry, and two people died.

In 1954, a very unusual case in England attracted a great deal of attention. Arthur Ford, the 44-year-old general manager of a wholesale chemist's developed a romantic obsession with 27-year-old Betty Grant, an attractive secretary with the company. His passion was not reciprocated.

On April 26, Ford stole some cantharidin from the company's stockroom. An extract of the dried beetle *Cantharis vesicatoria* (popularly known as "Spanish fly"), cantharidin has long enjoyed a false reputation as an aphrodisiac. It is, in reality, an irritant poison, the fatal dose being less than 60 milligrams.

The following day, Ford appeared in the office with a bag of coconut ice (candy), and offered pieces to Betty Grant and two other women. A little later, 21-year-old June Malins was also seen eating some (the third woman did not eat any). An hour later, June Malins complained of stomach pains, and Betty Grant took her to the sickroom; soon Grant herself became ill, and Ford looked pale and sick.

When a doctor arrived, he ordered all three to University College Hospital nearby. The two women grew worse, and died later that day, but Ford recovered. An autopsy on both women revealed that each had received a dose of 60–120 milligrams of cantharidin.

Ford was questioned by police, and confessed that he had doctored the coconut ice with the poison. At his trial in 1954, he pleaded guilty to a charge of manslaughter and was sentenced to five years' imprisonment.

CASE CLOSED

in 1828. Compounds of lead, mercury and antimony were prepared in pure form, and chemicals such as chloroform and ether were synthesized. All these, and more, found their way into the hands of poisoners.

Chloroform has a distinctive odour, and there are several relatively simple chemical tests for its presence, making it easy to detect as a poison. Many other compounds, particularly those isolated from natural sources, are very difficult to identify. Some early toxicologists had to rely upon their sense of taste – a remarkably dangerous proceeding.

When the Comte de Bocarmé and his wife were accused of poisoning her brother with nicotine in 1850, the Belgian chemist Jean Servais Stas made a liquid extract from the dead man's organs. When he tasted it, he reported a burning sensation on his tongue, mouth and throat. He subsequently validated his findings by killing two dogs with nicotine.

Giving evidence at a criminal trial later in the 19th century, Professor Robert Christison, then a leading English toxicologist, began to explain to the judge: "My Lord, there is but one deadly agent of this kind that we cannot satisfactorily trace in the human body after death, and that is —". At that point he was stopped, and prevented from naming the poison, which was aconitine. Extracts of the monkshood or wolfsbane plant *Aconitum napellus* were widely used as a poison in ancient times: the Greeks called it "stepmother's poison", and the emperor Trajan forbade Romans to grow wolfsbane in their gardens.

Aconitine's first effects are to produce a tingling sensation in the mouth. George Lamson was accused in 1881 of murdering his wife's young brother by giving him a slice of cake laced with the drug. Dr. Thomas Stevenson of Guy's Hospital, London, gave evidence on extracts he had obtained from the dead boy's organs: "Some of this extract I placed on my tongue," he said, "and it produced the effect of aconitia."

Aconitine – like nicotine, morphine, hyoscine, strychnine, and similar plant-derived drugs – is an alkaloid. Dr. Hawley Harvey Crippen used hyoscine to kill his wife in 1910. The symptoms of poisoning with many of these drugs are insidious, but the signs of strychnine poisoning are dramatically obvious. The muscles go into spasm, and breathing becomes difficult; the back arches in such a way that only the head and heels touch the surface on which the body is lying. The face becomes darkly suffused with blood, while spasms of the

Dr Crippen being escorted by Inspector Dew from the liner Montrose *on July 31, 1910. This was the first time in which radio was used in the apprehension of a murderer. The captain of the* Montrose, *his suspicions aroused by Crippen's lover, Ethel le Neve, who was dressed as a boy, sent a radio message to London. Dew boarded a faster ship, and overtook the fugitives off the Canadian coast.*

A contemporary newspaper illustration depicting the execution of Dr. Crippen at Pentonville jail in London.

mouth muscles produce a hideous grinning known as *risus sardonicus*.

Nevertheless, physicians have sometimes ignored these symptoms. When the notorious Dr. Thomas Neill Cream was poisoning prostitutes with strychnine, in London in 1891–92, the death of at least one of the girls was attributed to alcoholism. In May 1934, when Arthur Major died suddenly in Lincolnshire, England, his doctor gave the cause of death as "status epilepticus". It was only when the police received an anonymous letter, stating that a neighbour's dog had died after eating scraps put out by Arthur's wife Ethel, that they obtained an exhumation order and discovered that Major had died of strychnine poisoning.

The 19th century is famous in criminal history as the "golden age" of homicidal poisoning. It was, however, also a time when research chemists developed specific identification tests for the rapidly growing number of natural and synthetic poisonous substances, and when the first laws to control their availability were drawn up. As a result, during the 20th century – and particularly since the introduction of laws to control poisons – most murders by poison were committed either by medical practitioners or by those whose work brought them into contact with drugs or other toxic substances.

Deadly Medicine

Today, the development of drugs for the treatment of specific diseases has increased the number of available poisons to many thousands.

The most widespread group of drugs has been, until quite recently, the barbiturates. Adolf von Bayer first synthesized barbituric acid in 1863: he named it after his friend Barbara. About 40 years later, two derivatives were prepared and found to be highly effective sedatives. These were diethylbarbituric acid – known as barbitone, or veronal – and phenobarbitone, or luminal. The dangerously addictive nature of these substances was not recognized until the 1950s but by then many millions of tablets had been prescribed by doctors for their patients. Many accidental and suicidal deaths were attributed to barbiturate poisoning, but who knows how many of these were actually homicidal? Relatively few cases have been shown to be murder.

The recent history of homicidal poisoning shows the range of unexpected substances that can be used, and the difficulties that toxicologists face in identifying them. In 1978, for example, the Bulgarian refugee journalist Georgi

FACT FILE

Modern forensic laboratories use a wide range of sophisticated equipment, and many stages in the analysis of poisons can now be performed by computers. However, toxicologists have to have some idea of what they are looking for, before they can hope to identify a specific poison. If the victim has been seen by a physician before death, the clinical symptoms will suggest a possible type of poison. Alternatively, the pathologist's report on the autopsy can often indicate the probable cause of death.

Without this evidence, the toxicologist may have to resort, in the first place, to long-established chemical procedures. Gone, however, are the days when sizeable samples were necessary for chemical tests. There is now a large number of "spot tests" for specific poisons, which either produce a coloured reaction, or the formation of a precipitate or characteristic crystals, which can be further examined under a microscope.

Poisonous agents can be classified, broadly, into four types:

- Substances, mostly liquids, that are volatile in steam or a current of air or inert gas.
- Water-soluble substances, not volatile in steam, and not soluble in organic solvents.
- Organic substances, either natural or synthetic, that are more soluble in organic solvents than in water.
- Inorganic elements, mostly metallic, that are not volatile in steam. They should be looked for after all organic matter has been destroyed. In searching for poison, toxicologists generally find the greatest concentrations in the liver and spleen, and this is particularly true of metallic elements.

Making use of this somewhat crude classification, toxicologists now have a wide range of analytical techniques at their disposal. These include a number of spectroscopic methods of different types, liquid and gas chromatography, electrophoresis, mass spectrometry, neutron activation, "tagging" with radioactive tracers, and immunoassay. (See "The Forensic Hardware".)

Many functions can be performed by completely automated instruments, controlled by computer. This relieves the laboratory personnel of laborious repetitive work. Nevertheless, there are still cases in which laboratory mice must be used, to compare the effects of a suspected poison with a known substance.

Markov was assassinated in London, presumably by the Bulgarian Secret Service, with a pellet fired from a gas-powered gun hidden inside an umbrella. It contained ricin, an extract of castor oil with a fatal toxicity as low as 2 millionths of the body weight – making it twice as poisonous as cobra venom.

In June 1981, 28-year-old Susan Barber murdered her husband Michael by adding the herbicide paraquat to his steak and kidney pie. Barber's body was cremated, but nine months later – following a continuing inquiry into the cause of his death – serum and organ samples that had been saved from the autopsy revealed the presence of the herbicide. Mrs. Barber and her former lover Richard Collins were tried: they were both found guilty in November 1982. Mrs. Barber was sentenced to life, and Collins to two years' imprisonment.

CRIME FILE: Graham Young

A convicted teenage poisoner, supposedly rehabilitated, he quickly returned to his former obsessions. Two men died, and several more suffered the debilitating effects of thallium poisoning.

A still from the British film The Young Poisoner's Handbook, *which chronicled, with distinct black humour, the career of Graham Young, while catching, with grim accuracy, the dreary flipside of London's "swinging sixties".*

In 1962, an English 14-year-old named Graham Young admitted murdering his stepmother by poison, and attempting to poison his father, sister, and a schoolfriend. He was committed to Broadmoor Institution for the Criminally Insane, but in 1971 he was pronounced "cured", and released.

Young soon found employment as assistant storeman with a company manufacturing photographic lenses at Bovingdon, in Hertfordshire. By chance an outbreak of gastroenteritis had occurred in the vicinity at that time, and was nicknamed the "Bovingdon bug". When the company's foreman, Bob Egle, fell ill two months later, in July, it was assumed that he had been infected by the same "bug". He soon became worse, however, and died within a week or two. His cause of death was described as "peripheral neuritis". Egle's replacement, Fred Biggs, began to show similar symptoms in October, and died on November 19. Throughout the winter, other employees fell ill, and two in particular lost much of their hair.

The firm's managing director grew concerned that there might be a leakage of chemicals somewhere on the premises – a compound of thallium was one of those employed in lens manufacture. He called on the services of a team of toxicologists, led by Dr. Iain Anderson. When Dr. Anderson met the staff, he was surprised to be asked by Young whether the symptoms of the victims might be attributed to thallium poisoning.

After consulting the literature on the subject, Dr. Anderson was sufficiently concerned to ask Scotland Yard if they had any record on Graham Young, and was shocked to discover his history. Young was arrested on suspicion, and a search of his lodging uncovered a diary in which he had kept a meticulous record of his activities. It seemed he had been administering thallium in cups of tea he brought to fellow employees. Of Fred Biggs he wrote: "have administered a fatal dose of the special compound to F .... I gave him three separate doses."

Analysis revealed the presence of thallium in the victims. Although Bob Egle had been cremated, an analysis of his ashes showed the presence of a remaining nine milligrams of the metal. In addition, the police found a packet of the thallium compound in the lining of Young's coat, which he told them had been intended as his "exit dose". He was found guilty in June 1972, and committed to life imprisonment.

CASE CLOSED

One of the most horrific cases of recent years was that of Genene Jones, a hard-working children's nurse in San Antonio, Texas. Jones was probably responsible for the death of more than 30 babies and youngsters in her care. She derived such excitement and pleasure from resuscitating and caring for babies who suffered sudden cardiac arrest that she began to inject them with near-lethal drugs to experience the thrill of nursing them.

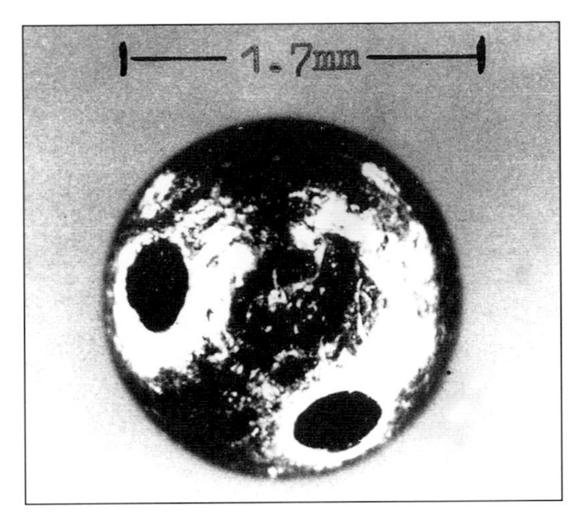

The platinum/iridium pellet that was fired from an umbrella gun into Georgi Markov's thigh in 1978 measured less than 1/12 in (2mm) in diameter, and contained 0.2 mg of the deadly poison ricin.

The drug that Genene Jones favoured was succinylcholine, popularly known as "synthetic curare". In the end, the discovery of vials that she had tampered with led to her detection. Succinylcholine, like all curare-related synthetics, relaxes and paralyzes muscle fibre, resulting in the inability to breathe. Jones was found guilty of murder on February 15, 1984, and sentenced to at least 25 years' imprisonment before her first chance of parole.

The psychological condition from which she suffered has been named Munchausen's syndrome "by proxy" – several similar cases have been recorded. In 1989, another American, male nurse Richard Angelo, was found guilty of injecting four patients with muscle-relaxing drugs. In a videotaped confession he said: "I felt inadequate .... I felt I had to prove myself." In 1991, in England, Nurse Beverley Allitt was found guilty of four murders and three charges of attempted murder on children in her care. She had injected many of them with massive doses of insulin.

Genene Jones, a children's nurse at the Medical Center Hospital in San Antonio, Texas. She caused the deaths of probably more than 30 children in her care by administering fatal doses of succinylcholine.

Heavy Metal

Although most modern poisoners rely upon complex organic drugs of one kind or another, many inorganic (mineral) substances are equally poisonous – they may frequently be slower-acting, but their effects are as lethal, and sometimes more agonizing.

Arsenious oxide – or arsenic – is the member of this group that has been used most often with homicidal intent, but there are other, equally poisonous,

Richard Angelo (left), a male nurse at the Good Samaritan hospital, West Islip, New York, who was found guilty of murdering four patients in his care. "I felt I had to prove myself," he said.

compounds. All compounds of lead and mercury, for example, are toxic. The mercuric salts once used to cure felt for hats drove many workers insane – giving rise to the expression "As mad as a hatter".

The use of tartar emetic (potassium antimonyl sulphate) in the Bravo case has already been described in "Suicide or Murder?" Antimony – among other substances – was also the poison employed by Dr. William Palmer, who murdered possibly 14 people in England in the years before 1856. In an unusual case, French pharmacist Pierre-Desiré Moreau poisoned his second wife with copper sulphate in 1874.

Compounds of the metal thallium, which have been used in insecticides, rodenticides and depilatories, have also featured in homicidal poisonings. In 1949, in Sydney, Australia, middle-aged Caroline Grills was accused of four murders and two attempted murders with thallium. The prosecution pursued only one charge of attempted murder, and Caroline Grills spent the rest of her life in prison, where she was known to fellow-inmates as "Aunt Thally".

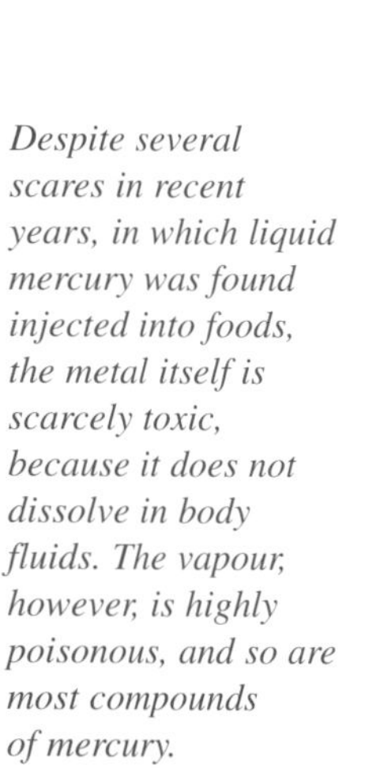

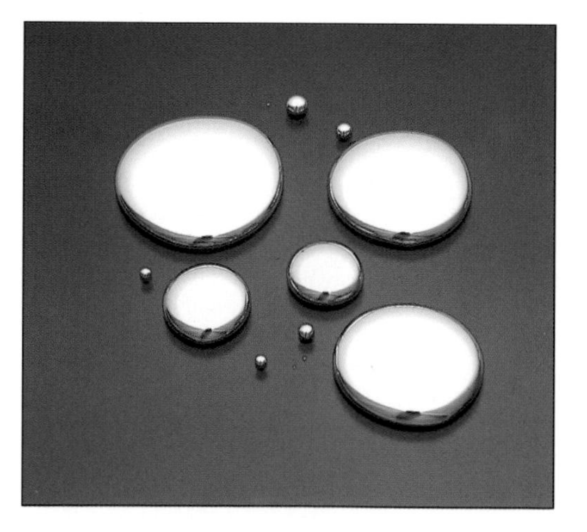

Despite several scares in recent years, in which liquid mercury was found injected into foods, the metal itself is scarcely toxic, because it does not dissolve in body fluids. The vapour, however, is highly poisonous, and so are most compounds of mercury.

Although not a metal, phosphorus has also featured in several cases of poisoning by inorganic substances. The element occurs in two forms: relatively inert red phosphorus, and the intensely poisonous white form, which has been used as a rat poison. Phosphorus was used in the murder of Sarah Ricketts, an elderly widow who lived in a bungalow in the seaside resort of Blackpool, Lancashire, on the northwest English coast.

On March 12, 1953, Louisa Merrifield and her husband Alfred became housekeepers to Sarah Ricketts. Very soon Louisa Merrifield was boasting that she had worked for an old woman who had died and left her a bungalow. Questioned further, she replied: “She’s not dead yet, but she soon will be”. On April 9, she called in Ricketts’s doctor to certify that the old lady was well enough to make a new will.

The trial of Dr. William Palmer in 1856. He was found guilty of poisoning John Cook with antimony, and was probably responsible for 13 previous deaths.

On April 14, Sarah Ricketts died. An autopsy revealed a dark brown liquid in her stomach that was a mixture of brandy and phosphorus. Although no traces of rat poison were found in the house, it emerged that Alfred Merrifield had previously bought a tin.

Husband and wife were arrested, and at their trial in July Louisa Merrifield was found guilty of murder, and sentenced to hang. The jury, however, were unable to reach a verdict on Alfred Merrifield.

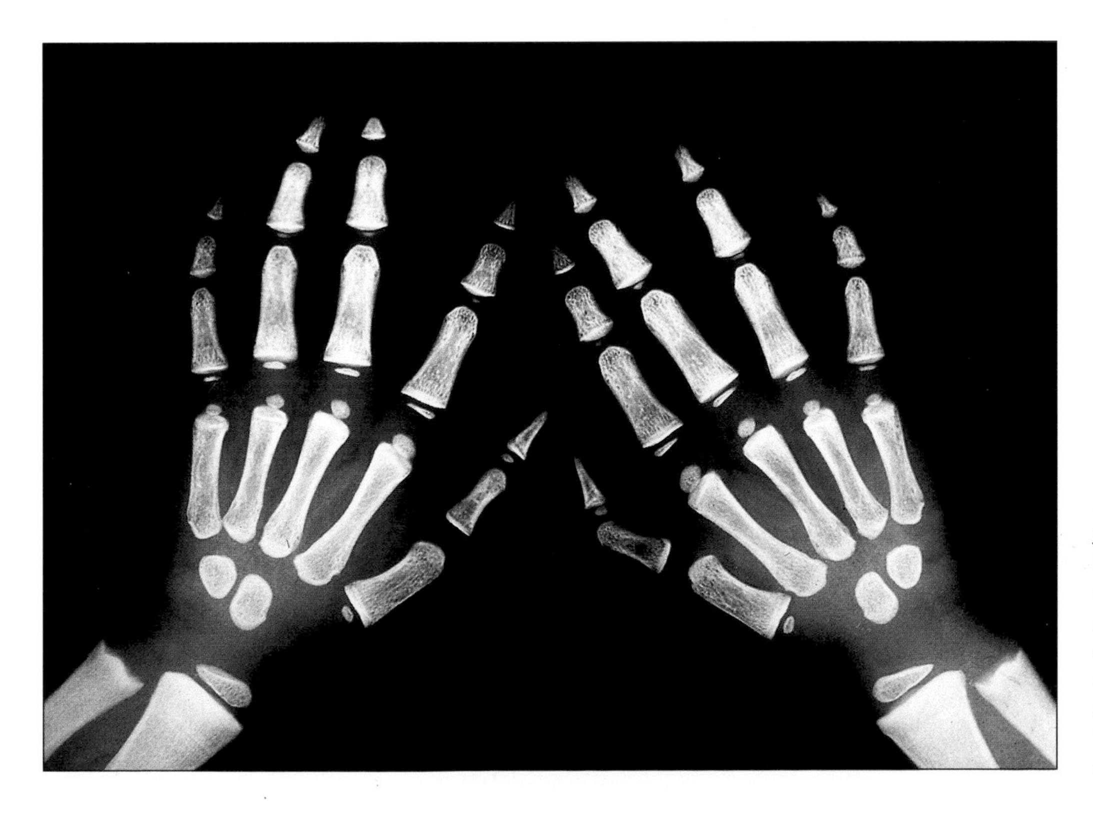

In adults suffering from lead poisoning, more than 95 percent of the element is concentrated in the bones. Using an iodine isotope as tracer, X-ray analysis will detect deposited lead.

Opposite page: The arrest of serial murderer John Christie in London in 1953. Timothy Evans had previously been found guilty, and hanged, for the death of at least one of his victims, Evans's wife.

Poisonous Gases

Most poisons are solids or liquids, but there are a number of poisonous gases. Principal among these are carbon monoxide and hydrogen cyanide.

Carbon monoxide is highly toxic because it has a strong "affinity" for the hemoglobin in the blood – some 300 times greater than oxygen. Essential oxygen is carried in the circulating blood by hemoglobin to all the body's tissues; if the oxygen is replaced by carbon monoxide, asphyxia rapidly results.

In the past, there was a ready source of carbon monoxide in the form of coal gas. For more than a century it was a major means of suicide; it was also the cause of many accidental deaths and was employed in a number of murders. The most notorious case of the homicidal use of coal gas was that of John Christie, who admitted to murdering his wife, and at least five other women, in London, between 1943 and 1949. Four were prostitutes, whom he rendered unconscious with gas, before strangling and raping them.

The replacement of coal gas by natural gas has removed the major danger, but incomplete combustion of all hydrocarbons, and of carbon in coal, coke or charcoal, also results in the production of carbon monoxide. Inefficient burners of natural gas, or of butane or propane, will therefore produce carbon monoxide, and car exhausts (unless fitted with modern catalytic converters) emit four to eight percent carbon monoxide, diesel engines rather less. Since pure carbon monoxide is odourless, it remains a potential cause of death unless proper precautions are taken to ventilate any enclosed space in which combustion takes place. In addition, in many deaths by fire the primary cause of death is inhalation of carbon monoxide.

Even small concentrations of carbon monoxide will displace oxygen from the red blood cells and rapidly diminish the ability of the blood to transport oxygen. There is wide variation in personal susceptibility and in the speed with which carbon monoxide replaces oxygen, but a replacement of some 50–60 percent is likely to be fatal in healthy adults; old people and those with pulmonary or cardiac disease may succumb at levels as low as 25 percent.

It is estimated that the blood of a healthy sedentary person breathing air containing as little as 1 percent carbon monoxide would reach 50 percent replacement in 15 minutes, but if the individual is moving actively this can be attained in five minutes. A concentration as low as 0.2 percent has been known to cause death within a few minutes.

Because the combination of haemoglobin with carbon monoxide continues to build up in the blood, even as little as 0.1 percent of the gas in the atmosphere can produce a fatal level in the blood within two to three hours. A car engine running in a closed garage can prove lethal within five minutes.

The symptoms of carbon monoxide poisoning are insidious, and the victim may be unaware of anything except a slight headache before lapsing into

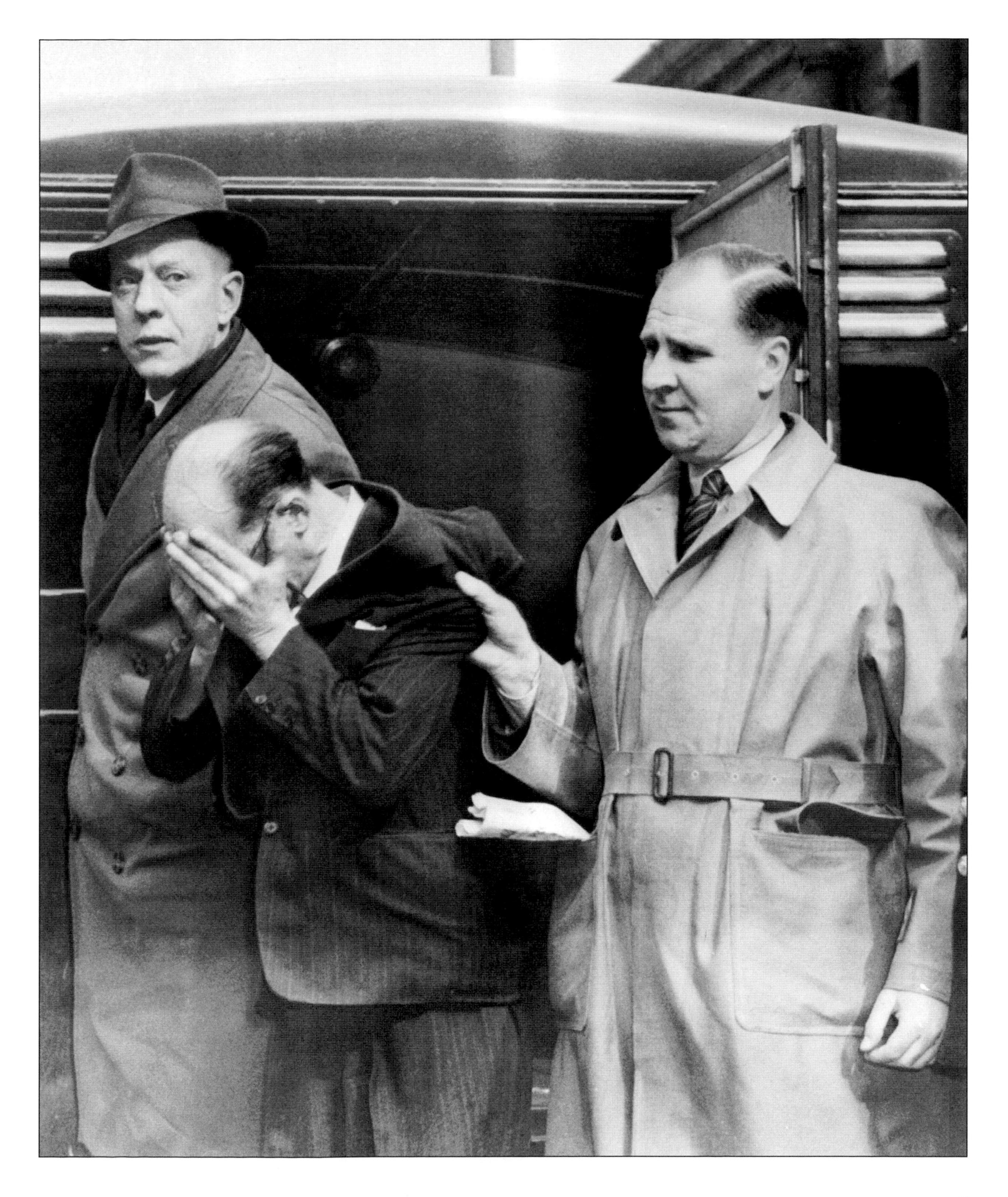

coma and death. At blood levels up to 30 percent, fit adults may experience no more than headache and nausea, and a loss of concentration that is sometimes mistaken for drunkenness. Levels of 30–40 percent are characterized by faintness, possible vomiting, blurring of vision, and a gradual slide into coma. Above 50 percent, death is almost inevitable, although healthy young adults have reached 70 percent before dying.

An infallible sign of carbon monoxide poisoning is the cherry-pink coloration of the skin, lips and internal organs, which can persist for many months after death. This colour is particularly noticeable in the hypostasis of those parts of the body to which the blood migrates after death. Confirmation of carbon monoxide in the blood is made either by spectroscopy or by gas chromatography.

Cyanide, like carbon monoxide, poisons by preventing oxygen reaching the body tissues, but in this case it acts by inhibiting the enzyme that effects the uptake of oxygen from the blood.

Hydrogen cyanide is a gas, while hydrocyanic (prussic) acid is a solution of the gas in water. The salts of the acid, usually sodium or potassium, are white solids. Cyanides are used quite widely in industry, in photography and electroplating; as rat and wasp killers; and in the fumigation of trees, fruit and ships' holds. Laboratory workers have ready access to cyanides, and many cases of suicide have occurred. In one survey of cyanide deaths, 70 percent were found to be suicidal, and the rest accidental, although cyanide has also been employed homicidally. Because of the minute quantity required to cause death, and because antidotes must be given at once and are seldom readily available, cyanide was the favourite poison in "suicide capsules" issued to spies and others whose interrogation after capture could be disastrous.

Hydrogen cyanide may also be produced in cases of fire, by decomposition of the structural foam plastics that are used in upholstering furniture.

A concentration of the gas in air as low as 0.02 percent can cause death almost instantly, and lower concentrations will prove fatal if inhaled over a period of time. As little as 50 milligrams of prussic acid can be lethal if swallowed, and the

FACT FILE

Although prussic acid is extremely poisonous, it can deteriorate over a period of time. All chemistry students know the story of the laboratory staff who were troubled by the persistent attentions of an unhealthy, mangy, stray cat. Eventually they prepared a dish of meat, and emptied into it most of the contents of a half-empty bottle of prussic acid that had stood for a long time on the shelf. The cat ate the food with relish, and left. Two or three days later it returned, a picture of health with a fine glossy coat. Subsequent analysis showed that the prussic acid had combined with atmospheric carbon dioxide to form ammonium carbonate, which had acted as a harmless laxative.

fatal dose of potassium cyanide is generally given as 250 milligrams – although there have been cases of recovery from doses as high as ten times this amount.

Hydrogen cyanide has a characteristic smell of bitter almonds, although it is said that up to 20 percent of the population are unable to detect it – one famous British pathologist could smell it only while smoking. Cyanide also occurs naturally, in the fruit kernels of almonds, for example, and in laurel leaves. Butterfly collectors have traditionally used crushed laurel leaves in their killing-bottles, and almonds may contain up to 0.1 percent cyanide. Only recently, the death was reported of a young woman who had eaten green almonds, and there is also the case of death following drinking from an old bottle of almond liqueur, in which the cyanide had concentrated in an oily layer. Undoubtedly the most obscene use of cyanide compounds was the Zyklon B used in the SS extermination camps. A more recent case is the "Jonesville massacre" in Guyana. On the orders of their leader, "Reverend" Jim Jones, 900 members of the People's Church gave cyanide-laced Koolaid to their children, then queued up to receive cyanide injections.

Japanese police evacuate the Tokyo subway, following the release of saran gas by the Aum Shirikyo sect on March 19, 1995.

The signs of death in cyanide poisoning are similar to those of carbon monoxide poisoning, since the cause is the same – oxygen starvation of the tissues. The bright pink colour of the blood is obvious in the skin, lips and internal organs, and shows up in the hypostasis. It can be mistaken for the coloration due to carbon monoxide, but is generally slightly darker. At autopsy, the characteristic smell of bitter almonds in the organs is usually obvious, and the presence of cyanide is easily established by chemical analysis.

Modern developments in chemical warfare have resulted in the preparation of a range of "nerve gases". These act by blocking the transmission of nervous

Arthur Waite awaiting trial in New York in 1916. He was charged with the murder of his mother-in-law and father-in-law, making use of bacterial cultures, and was sent to the electric chair in Sing Sing in May 1917.

impulses from the brain to the rest of the body, and can quickly cause death. Fortunately they have not so far been employed in warfare, but on March 19, 1995 members of the Aum Shirikyo sect released the nerve gas sarin in three stations of the Tokyo subway. The attack killed 12 people, and the police reported that more than 5000 had suffered ill effects.

Biological Agents

Probably the rarest form of homicidal poisoning is by bacterial cultures. In 1910, Dr. Bennett Clarke Hyde was accused of murdering the Kansas City millionaire, Thomas Swope, and the administrator of his estate, James Hunton, with a mixture of strychnine and cyanide. Shortly after the two men died, five beneficiaries of Swope's estate had been taken ill with typhoid, and one had died. At Hyde's trial, a bacteriologist testified that he had supplied him with typhoid cultures. There was strong suspicion that Hyde had injected his victims with the cultures, although it was impossible to prove. Hyde was found guilty of first-degree murder, but after numerous appeals a new trial was ordered in 1911. This ended as a mistrial, and a third ended with a hung jury. In 1917 Hyde was sent for trial again, but his lawyers pointed out that this was contrary to existing laws, and he was released.

Some years later, Hyde's wife, who had stood by him all along, separated from him. She had complained of a stomach ache, and he had offered to treat her. No doubt she preferred to consult another doctor, and remain alive.

A near-contemporary of Dr. Hyde was Arthur Waite, an ambitious New York dentist. In addition to his dental practice, Waite carried out bacterial research at Cornell Medical School. He admitted to killing his mother-in-law in 1916 by adding diphtheria and influenza cultures to her food. He also tried similar methods with his father-in-law: "Once I gave him a nasal spray backed with tuberculous bacteria". When the old man did not succumb, Waite finally murdered him with arsenic.

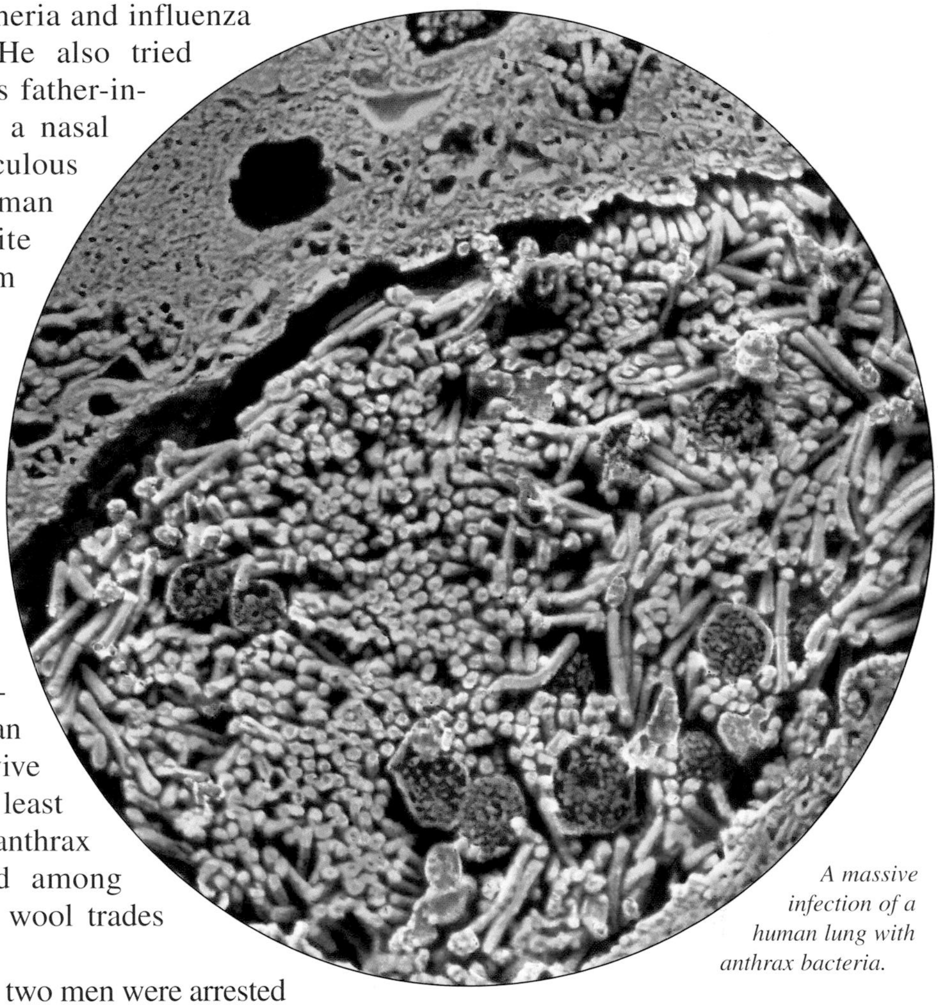

Among the biological warfare agents, one of the most feared is anthrax. Many servicemen in the Gulf War were inoculated with anthrax vaccine in case it might be used against them. Anthrax is a bacillus that normally kills cattle and sheep, but it can form spores that will survive in dry conditions for at least 40 years. Fatal cases of anthrax infection have occurred among workers in the hide and wool trades for this reason.

A massive infection of a human lung with anthrax bacteria.

On February 19, 1998, two men were arrested at Henderson, near Las Vegas, Nevada, allegedly carrying a container of anthrax culture. The FBI initiated a major alert when it was learnt that one of the men had previously been convicted, in 1995, of receiving three vials of bubonic plague culture through the mail. It was reported that he had spoken, in 1997, of placing a glass globe containing anthrax on the New York subway, and causing "economic ruin". Decontamination experts were happy to declare an "all clear" on the area, and in the biochemical laboratory where the men had worked.

A second anthrax scare occurred just over a year later. On February 23, 1998, pro-choice organizations in New York, Kansas City, Pittsburgh, and Delaware, among others, received letters containing a brown powder that, the letters said, was anthrax spores. The campaign was declared a hoax, but the ease with which some people are able to obtain bacteria, and culture them, has fuelled fears that terrorists could strike, silently and fatally, at any time.

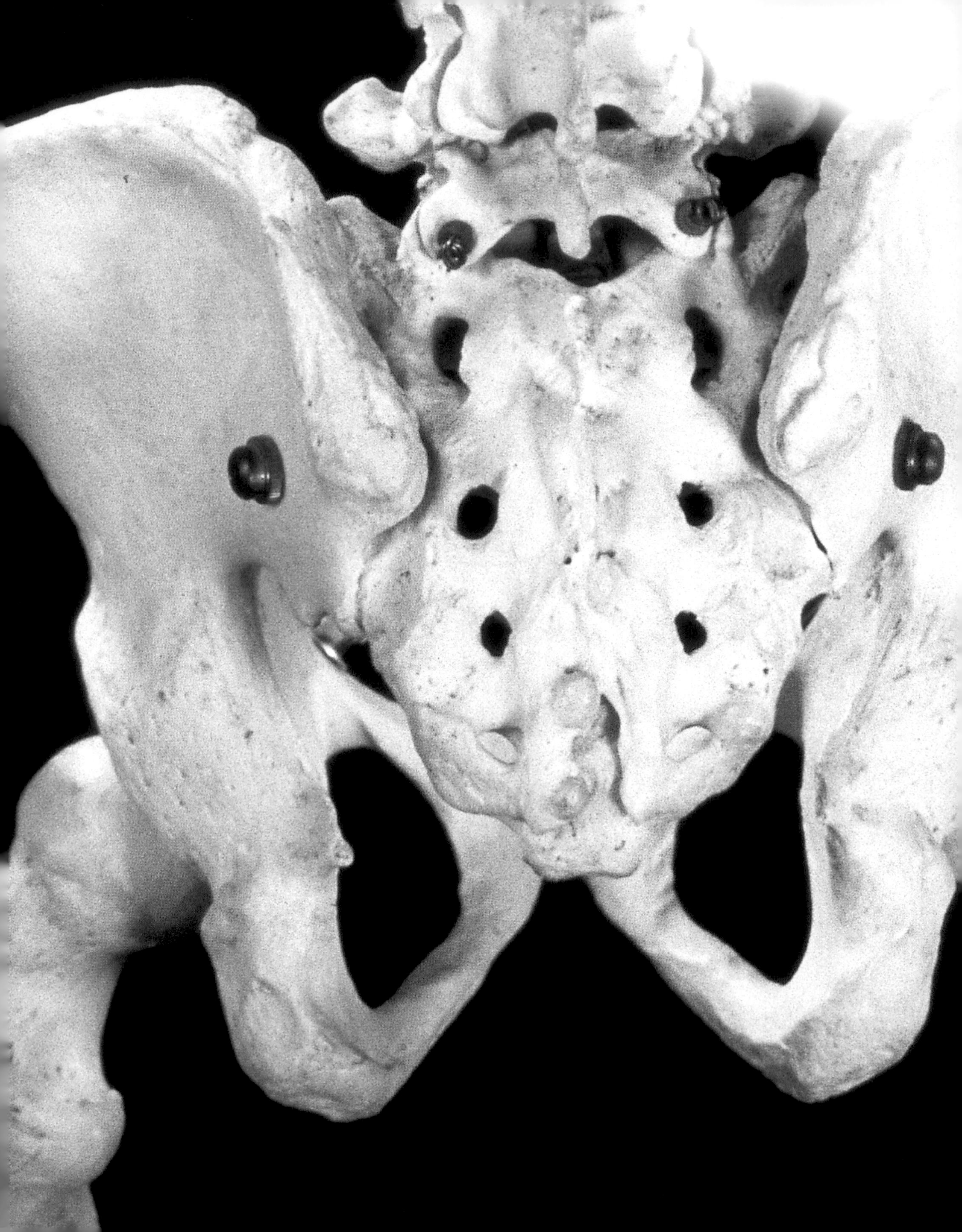

Skull and Bones

A close-up of a female human pelvis. This part of the skeleton provides an invaluable indication in determining the sex. Notice the characteristically broad spread of the pelvic cavity.

FROM THE MOMENT OF DEATH, the body begins to putrefy. Within days, the facial features may be almost unrecognizable; within weeks – under normal conditions – the tissues will liquefy; and gradually, all that is left is "a rag and a bone and a hank of hair." Each stage presents the investigator with increasing difficulties of identification. If the body has lain untouched in one location, even a bare skeleton can provide the pathologist with a number of clues. The task grows much more complicated if the body has been dismembered – whether by a psychopathic killer, or a disaster such as an explosion or an aircraft crash. This is particularly true if a number of bodies are involved.

SEX, AGE AND STATURE

Unquestionably, the most valuable indicators of the sex of a skeleton are the pelvis and the skull. The female pelvis, designed for the bearing of children, is broader and shallower than the male. The pelvic cavity is noticeably larger: as a rough guide, one pathologist has given its diameter in an adult woman as the spread of a thumb and forefinger, while the male cavity diameter is the spread of a forefinger and middle finger. There are also differences in the size and shape of other pelvic bones.

Male characteristics in the skull only begin to develop after the age of 14, and sexing the skull of a younger person is difficult and unreliable. However, there are distinct differences between male and female adult skulls. In general, the orbit (eye socket) is rounded in females, but much more rectangular in males. The nasal aperture of a male is longer and narrower, shaped almost like a teardrop, while a woman's is more pear-shaped. The female jaw is rounded, the male jaw more angular, and generally larger and heavier. In addition, a woman's forehead does not slope back so much as a male's, and usually lacks a pronounced brow ridge over the eyes.

There are other reported differences between male and female bones: those of males will probably be heavier, for instance. From a forensic point of view other differences, however, can only be taken in conjunction with the more obvious sex characteristics.

To determine the age of a skeleton, the pathologist examines the skull and individual bones: this can usually be carried out even when only a few bones are available. In a new-born infant, the ends of the long bones are attached to the main shaft by cartilage – known as an epiphysis. Gradually this attachment

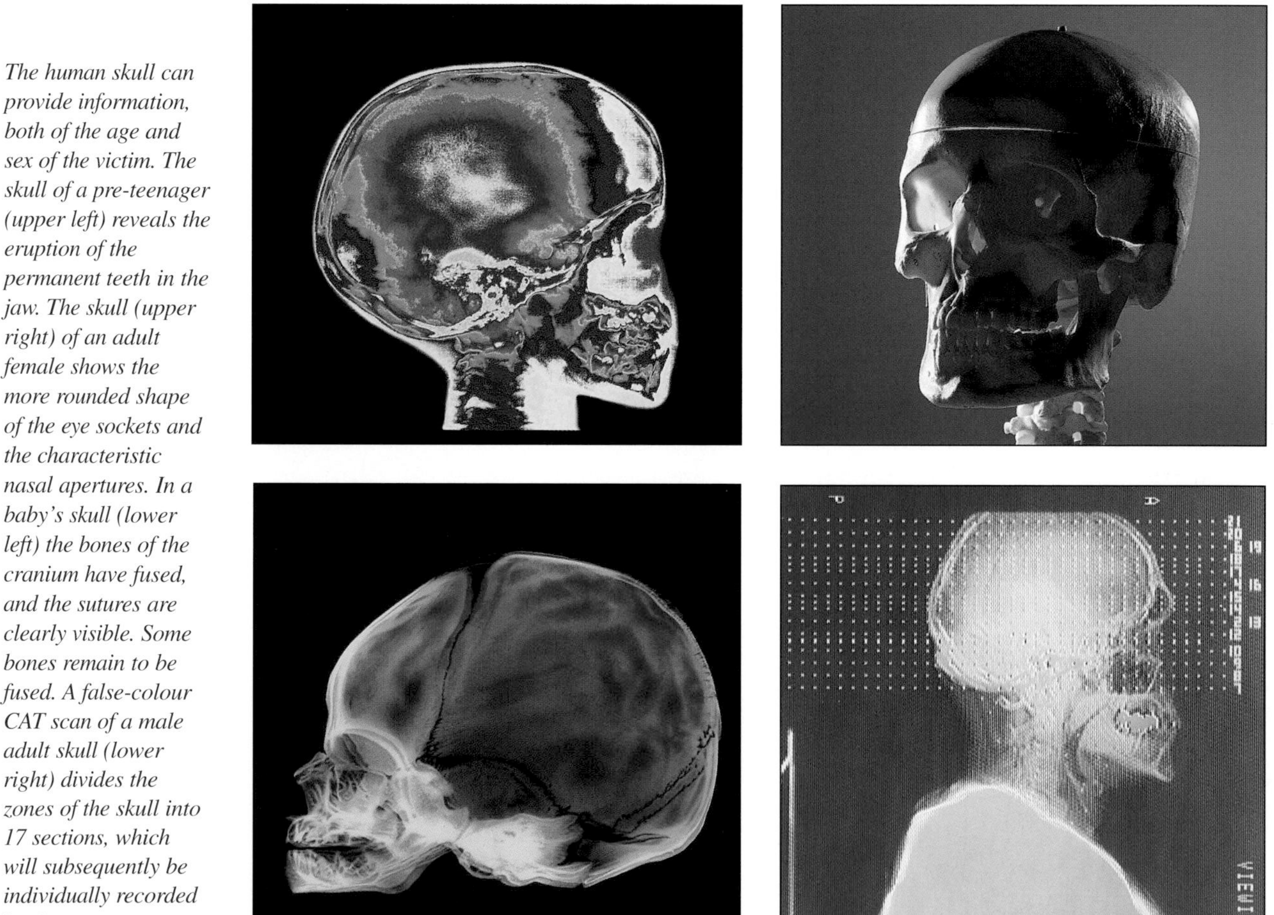

The human skull can provide information, both of the age and sex of the victim. The skull of a pre-teenager (upper left) reveals the eruption of the permanent teeth in the jaw. The skull (upper right) of an adult female shows the more rounded shape of the eye sockets and the characteristic nasal apertures. In a baby's skull (lower left) the bones of the cranium have fused, and the sutures are clearly visible. Some bones remain to be fused. A false-colour CAT scan of a male adult skull (lower right) divides the zones of the skull into 17 sections, which will subsequently be individually recorded by the scanner.

disappears, and the two pieces of bone fuse together. This growth process can continue up to the age of 30 or thereabouts, and is detected either by direct examination or from an X-ray. The different stages of fusion can give an indication of age that is accurate within two or three years. After the age of 30, however, detectable changes in the bones cannot provide an estimate closer than some ten years.

In a similar way, the infant skull is in a number of pieces, marked by "sutures", which close up in stages. The frontal suture is the first to close, usually early in life. Other sutures normally begin to close between the ages of 20 and 30, but some can remain open or only partially closed to the age of 60. The last does not close until age 70 or later.

It is clear, then, that estimates of age based only on the bone structure can be fairly accurate until the mid-20s, but become progressively imprecise in later life, so that other factors must also be taken into consideration.

In calculating the stature, the bones are laid out on a special "ostiometric" board, which allows more accurate measurements than can be made with a tape and callipers. A reasonably correct measurement of height can be made when a major part of the skeleton is available, but there are also rules that allow an estimate of stature to be calculated from the long bones alone. These were first laid down by the French pathologist Rollet in 1888 and, although they have been superseded by more recent research, they have not changed much in principle. As a rough guide, the length of the humerus (upper arm bone) is 20 percent of the stature; the femur (thigh bone) is 27 percent; the tibia (lower leg bone) is 22 percent; and the spine is 35 percent. Nevertheless, all sorts of allowances have to be made for sex, age and race. Despite claims that racial type can be determined from variations in these ratios, the specific shape of the skull and pelvis are better guides in this instance.

Some remarkable deductions have been made from deformities of the skeleton. As a young pathologist in Cairo, Sir Sydney Smith was asked to examine a small parcel, containing three pieces of bone that had been found in a well. He identified three parts of a female pelvis: the two hip bones and the sacrum. They were small, and could have been those of a young girl, but the near-fusion of the crest of the hip bones indicated an age of between 22 and 25, while grooving of the bones showed that she had had at least one pregnancy. The right hip bone was bigger and heavier than the left, and the socket for the femur was also bigger. Fragments of tissue still adhered to the bones, and there was a lead pellet embedded in the right hip.

From these few observations, Smith was able to tell the police: "They are the bones of a young woman. She was short and slim, aged between 23 and 25 when she died, which was at least three months ago ... Her left leg was

CRIME FILE:

George Shotton

Forty years were to pass before the remains of the bigamist's wife were discovered. Meticulous forensic reconstruction eventually established his guilt – but it was too late.

Mamie Stuart with George Shotton, who married her bigamously. When she disappeared, he was the prime suspect.

In 1961, three young potholers exploring a cave at Caswell Bay near Swansea in Wales found skeletal remains concealed behind a large rock. The bones were taken to the Home Office Forensic Laboratory in Cardiff, where they were found to constitute a nearly complete skeleton. The body had been sawn into three pieces of roughly equal length: through the lower part of each thigh, and across both upper arms, the shoulder blades and spine.

The skull and pelvis suggested that the bones were those almost certainly those of a young female. Her height was estimated at about 5 feet 4 inches (1.6 metres), both from the appearance of the assembled skeleton and from measurement of the long bones. X-ray examination established that they were those of a mature adult, but that full maturity had only just been reached. The wisdom teeth in the jaw suggested that the woman was over 20. Furthermore, 2 bones at the base of the skull had only recently fused, suggesting that she was no more than 28 years old.

Some disintegrated scraps of clothing and sacking were found with the body, as well as items of jewellery and a celluloid hair-grip, still holding a few strands of brown

The "murder bag" belonging to the forensic expert at the 1961 recovery of Mamie Stuart's remains. Although more modern equipment is now used, the principles of evidence gathering remain the same.

hair. There was also a wedding ring, whose hallmark dated it to 1918. Finally, some gilt tassels were identified as probably coming from a style of stole fashionable in the early 1920s.

It was clear that the woman had died in suspicious circumstances and been disposed of surreptitiously. The crime had occurred about 40 years before, and presented a thorny problem for the police. Many records had been destroyed by bombing during World War II, but inquiries discovered several people who remembered the mysterious disappearance of former chorus girl Mamie Stuart in 1919–20. Newspaper files of the time soon turned up more information.

In 1918, a marine engineer named George Shotton went through a form of marriage with Mamie Stuart in north-east England, although he already had a wife and child living in south Wales. Eventually the couple settled near Swansea in November 1919. The last Stuart's parents heard of their daughter was a Christmas telegram wishing them "the compliments of the season."

In March 1920, the manager of a hotel in Swansea handed over to the police a suitcase that had been left unclaimed. It contained items of women's clothing and a scrap of paper with the address of the parents, who identified the clothes as having belonged to their daughter. The police issued a description, which read: "age 26; very attractive appearance; height five feet three or four inches; well-built; profusion of dark brown hair, worn bobbed ..." – but there was no trace of the missing woman, even when the police made an exhaustive search of the couple's home and the surrounding grounds.

George Shotton was soon found, living with his real wife and child near Caswell Bay. In May 1920 he was charged with bigamy, and at his trial the prosecuting counsel accused him of Stuart's death, but without a body nothing could be proved. He was found guilty of bigamy, and served 18 months' hard labour.

Over the years, newspapers regularly revived the "mystery of the missing chorus girl", but it was 40 years before the mystery was solved. The inquest, held in December 1961, was unusual, in that the skeleton of Mamie Stuart was laid out on a table in the well of the court. The jury returned a verdict of murder, naming George Shotton. The police soon found him – in a Bristol cemetery. He had died of natural causes in 1958, at the age of 78.

Two crime squad detectives and a forensic expert, with one of the sacks in which Mamie Stuart's dismembered body was wrapped.

CRIME FILE:

Dr. John White Webster

He disappeared in 1849. It was the discovery of his dentures that led to the identification of his remains, and the condemnation of his murderer.

The identification of the remains of Dr. George Parkman in 1849 was the first case in which dental evidence was to prove important.

Dr. Parkman was a wealthy and influential member of society in Boston, Massachusetts. He had endowed the chair of anatomy and physiology at Harvard University Medical College, and the new laboratories of the professor of chemistry, Dr. John White Webster. Over several years, Webster had borrowed money from Parkman, and the two men made an appointment to discuss repayment on November 22. Parkman did not return from the meeting, and was never seen again.

A college janitor, Ephraim Littlefield, recalled that, after Parkman's disappearance, Professor Webster's laboratory door had been locked, but the wall backing on to the laboratory's furnace had been very hot. During the Thanksgiving Day holiday, he took a hammer and cold chisel and broke through into a small vault; peering through the hole into the gloom, he saw a pelvis and two parts of a leg. When the police arrived, they found the upper parts of a human torso in a tea chest, and in the furnace they discovered fragments of bone – as well as a set of dentures.

In custody, Webster attempted suicide with strychnine, but the dose was insufficient, and he recovered. More than 150 human fragments were gathered from his laboratory, and a team of Webster's colleagues undertook the task of identification. They concluded that the remains were those of a male, some 5 feet 10 inches (1.75 metres) tall, and aged between 50 and 60. Parkman had been 5 feet 11 inches (1.78 metres) tall, and his age was 60.

At Webster's trial for murder, the clinching evidence came from Dr. Nathan Keep, a Boston dentist. About three years previously, Parkman had come to him for a set of dentures and, because his jaw was unusually prominent, Keep had made and kept a special cast. In court, he demonstrated how the dentures found in the furnace fitted the cast. Moreover, Parkman had complained that they made his mouth sore, and Keep had filed them down. He was able to point out the marks of his file on the dentures.

Webster was found guilty, and made a confession before his execution. A bitter argument had broken out between the two men, Parkman shouting "I got you your professorship, and I'll get you out of it!" Webster had picked up a block of wood, and struck him dead. Then he cut up the body, burning some in the furnace, and storing the larger pieces for later disposal.

CASE CLOSED

shorter than her right, and she walked with a pronounced limp. Probably she had polio when a child. She was killed with a shotgun … from a range of about three yards …" Armed with this description, the police soon learnt of a missing woman, short and slim and aged 24, who walked with a limp. She had been married, bearing one child, and then divorced, going to live with her father. He confessed that he had accidentally shot his daughter while cleaning his gun. He said that he had nursed her for a week until she died, and then had disposed of her body.

Teeth

After 50 years, skeletal bones have lost half their nitrogen content, and gradually become lighter and crumbly, until they are of little use to the forensic scientist. The teeth, however, can survive much longer, even enduring extreme conditions such as fire. They can provide a measure of the age of the body from which they have come as well as an important means of identification.

The assessment of age is easiest in relatively young people. The stage of development of first or second teeth can give a fairly accurate indication, although there can be considerable variation. X-rays of the jaw can reveal other teeth still developing. The third molars, or "wisdom" teeth, do not usually emerge until the late teens or early 20s.

Once all the teeth have emerged, it is possible – as with horses and other animals – to make a rough estimate of age by noting the condition of the

The Noronic *ablaze in Toronto harbour on the night of September 7, 1949. Many passengers were trapped aboard, and eventually 118 badly burned bodies were recovered. After five months intensive investigation, all but three were identified, many from their dental records.*

CRIME FILE:

Ted Bundy

A notorious serial killer, Ted Bundy escaped from detention and continued his murderous career. But the identification of his teeth marks on the body of his last victim was the evidence that led him eventually to the electric chair.

Serial killer Theodore "Ted" Bundy, who may have murdered more than 40 victims. He was sent to the electric chair in January 1989.

A man broke into a sorority house at Florida State University, Tallahassee, in the early hours of January 15, 1978, and brutally attacked four women. Two were seriously injured, and two – Margaret Bowman and Lisa Levy – were killed. Bite marks were found on Levy's left buttock.

A month later, a man giving his name as Chris Hagen was arrested in Pensacola for a minor driving offence, and a check of the records revealed that he was Ted Bundy, a wanted felon and suspected serial killer.

Bundy's murderous career had begun in Seattle in 1974, and he had then moved on to Utah and Colorado. In August 1975 he was arrested in Salt Lake City, where he was studying law, and charged with murder, the hair of one of his victims having been found in his car. He was extradited to Colorado on further charges, and held in the prison at Aspen to await trial.

Bundy, much to the embarrassment of the authorities, succeeded in twice escaping from Aspen. On the first occasion, he was tracked down, within eight days, to a shack on a neighbouring mountain. But on the second, the night of December 30, 1977, he made his way, first to Chicago, then Ann Arbor, Michigan, on to Atlanta, finally arriving in Tallahassee in the second week of January 1978.

Police thought it probable that Bundy was the perpetrator of the FSU attack.

When he refused to provide a dental impression, detectives obtained a warrant, allowing them to use force if necessary, and Bundy reluctantly agreed. Odontologist Dr Richard Souviron took colour photographs of his uneven teeth, both from the front and from inside his mouth, before making a complete cast.

Bundy's trial for murder opened in Miami on June 25, 1979, with Bundy conducting his own defence. In evidence, Souviron placed an acetate overlay of Bundy's teeth over an enlarged photograph of the bite mark on Levy's buttock, and showed how the two fitted exactly. Dr Lowell Levine, chief consultant in odontology to the New York Medical Examiner, confirmed the findings, and Bundy was found guilty. Before he went to the electric chair in January 1989, he hinted that the total number of his victims was more than forty.

CASE CLOSED

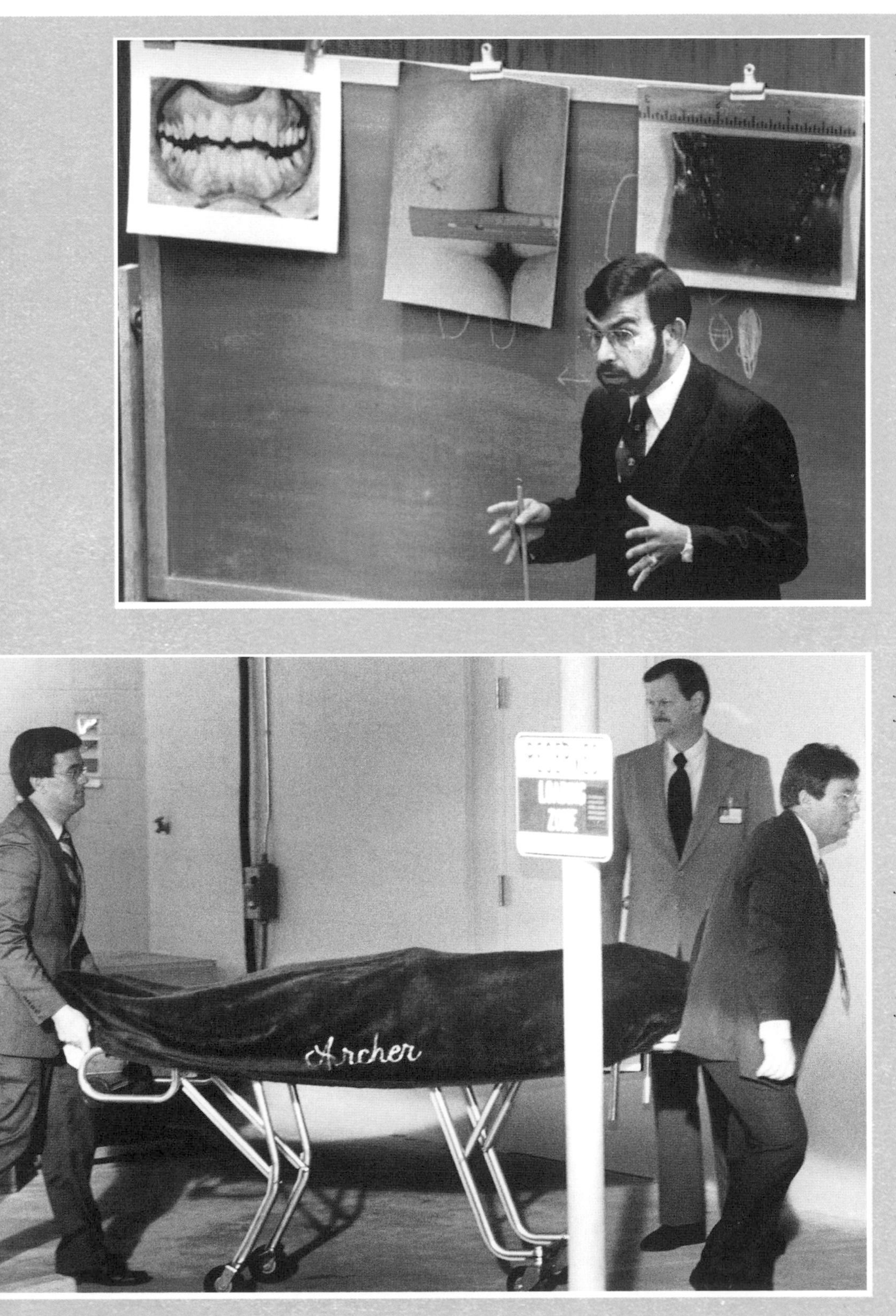

Dr Lowell Levine, New York forensic odontologist, testifies in court that the bite marks found on Lisa Levy's buttock matched the characteristics of Bundy's teeth. This evidence helped secure a guilty verdict and a sentence of death.

Bundy kept himself from the electric chair for a decade, entering one appeal after another, and maintaining his notoriety with innumerable interviews. He was finally executed in Florida State Prison on 24 January 1989. His last words were: "Give my love to my family and friends." Here, following his electrocution, his body is taken on a gurney to the Medical Examiner's office.

teeth, their degree of wear, the thickness of the dentine layer, and other indications. In the 1950s, the Swedish Professor Gosta Gustafson devised a six-point system for the recording of changes in these features. The advantage of this system is that it relies upon visual observation, and avoids any destructive interference with the evidence.

The examiner rates each degree of change on a scale of one to four. As an example, wear in one case was scored at 1.5, indicating an age between 14 and 22; the actual age was 18. In another example, total changes scored 12, which Gustafson estimated as indicating an age between 66 and 76; the actual age was 68.

Deducing the age of adult teeth is therefore a very approximate method, but identification of persons by their teeth, both real teeth and dentures, has proved itself many times. This can often be based upon known features, such as crooked, missing or gappy teeth, or markings due to the occupation or habits of the owner. Gustafson claimed, for example, to be able to distinguish a brass instrument player from a woodwind musician by the effect upon the teeth. Nonetheless, identification is made very much easier when dental records are available.

Nowadays, most people pay regular visits to the dentist, and detailed records are kept of fillings, extractions, bridges and dentures, plus any peculiarities or deformities. Throughout the world, there are about 200 different systems for making records, but they all provide a means of identification that is relatively easy, and almost 100 percent reliable. Dental records have proved invaluable in confirming the identity either of a single person (living or dead), or of a number of victims in a mass disaster. The technique has become a forensic speciality known as odontology.

In 1949, fire broke out aboard the cruise ship *Noronic* in Toronto harbour. When the blazing inferno was finally extinguished, 118 passengers – 41 male and 77 female – were dead. Of these, 20 were identified solely on dental evidence, it was a principal lead in another 20, and was important in the identification of a further 19. A fire in a hotel near Voss, in Norway, in 1959 claimed 24 victims. Of these, six were identified by dental evidence alone, and another nine by dental evidence and objects found by their bodies. (The victims included a number of Americans, and the existence of their X-rays, as well as dental records, made identification even easier.)

Computers were first used in the analysis of dental data for identification purposes in 1976, when 139 victims were recovered from the Big Thompson Canyon flood in Colorado. When American Airlines Flight 191 crashed at Chicago's O'Hare airport on May 25, 1979, Edward Pavlik, a local odontologist, headed a team of some 20 dentists who helped to identify 273 burnt and dismembered victims (see below). In 1981, the body of President Kennedy's assassin, Lee Harvey Oswald, was exhumed, to compare the

When Big Thompson Canyon flooded in Colorado in 1976, 139 people were killed. This was the first occasion on which computer analysis of dental data was used in identification.

teeth with his military records, and scotch prevalent rumours that he had been impersonated by a Russian spy.

From time to time, steps have been taken to include identification marks in teeth and dentures. During World War II, the Canadian Dental Corps incorporated a piece of nylon in acrylic dentures, to identify the soldier issued with them. In 1986, the American Dental Association inaugurated a programme to bond a coded micro-disk, little bigger than a pin head, on a patient's upper molar. This can provide a positive identification when read and matched by computer. It is hoped that, in due course, every American citizen may be coded in this way, as soon as the second teeth have emerged.

Bite Marks

To make a dental impression, the patient bites into a moulding material, from which casts can subsequently be made. In the same way, the marks made by biting teeth may equally provide important evidence in a criminal case. The first time that such evidence was allowed in a court of law was in 1906, when two English burglars were convicted. One had unwisely taken a

CRIME FILE:

Dr. Buck Ruxton

He murdered his wife and her maid, and attempted to make their mutilated bodies unidentifiable. However, painstaking anthropological reconstruction, and a pioneering use of photography, successfully established the identity of the victims.

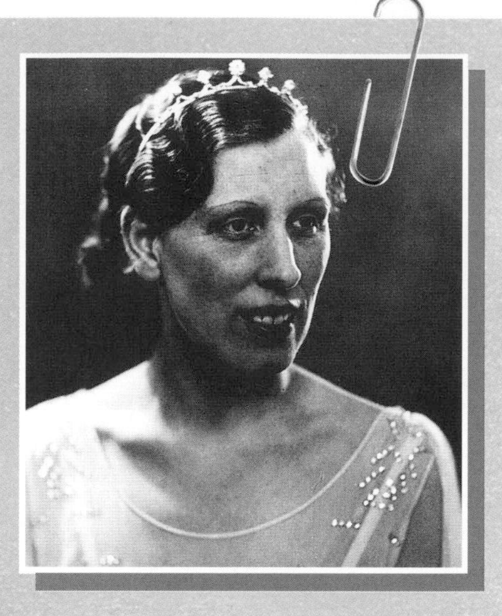

The identification of the scattered remains along the Carlisle–Edinburgh road was a triumph of the anatomist's skill. Professor James Brash introduced a new and immensely valuable technique. Working with a studio photograph of Mrs. Ruxton (above), he superimposed a photograph of one of the mutilated skulls (right). The exact match established that the skull was that of Isabella Ruxton.

On September 29, 1935, a woman crossing a bridge on the Carlisle-Edinburgh road near Moffat in Scotland saw a number of parcels, wrapped in newspaper and pieces of clothing, lying on the banks of the stream below. One, she realized, revealed a human arm. A police search of the area turned up some 70 human body parts; later, on October 28, a left foot was found wrapped in newspaper 9 miles south of Moffat, and on November 4 a right forearm and hand were found lying at the roadside not far from the bridge.

Two forensic experts examined the remains: Professor John Glaister, of Glasgow University's department of forensic medicine, and Professor James Brash of Edinburgh University's department of anatomy.

The various bundles found consisted of two heads; two portions of a single trunk; and two complete sets of limbs in fifteen pieces, with the exception of a missing right foot. The bodies had been neatly dismembered with a knife, and the murderer had removed deliberately most body parts that might lead to identification. The eyes were missing from one skull; a number of teeth had been extracted from the second, and the eyes, nose, ears, the tip of the tongue and the lips were missing. Most of the fingertips were missing, and the hands mutilated, but the right hand of one body still retained some visible prints. All the flesh on the legs of one body had been cut away, and among the many pieces of soft tissue found in the remains were three breasts and a uterus.

Re-assembly of the fragments required all of Professor Brash's anatomical knowledge, but eventually the investigators managed to put together two bodies that they identified as female.

Body No. 1 had no closed sutures in the skull, indicating an age under 30; and the epiphyses were not yet completely fused, so that the age range was set between 18 and 25. None of the wisdom teeth had yet emerged, indicating a probable age of between 18 and 21. Although the entire trunk was missing, the stature was calculated at between 4 feet 10 inches (1.45 metres) and 4 feet 11 inches (1.45 metres).

The skull sutures of body No. 2 suggested an age of 35 to 55 years, while the fusion of the epiphyses indicated a minimum age of 25. Development of osteoarthritis in the spine and right hip was a

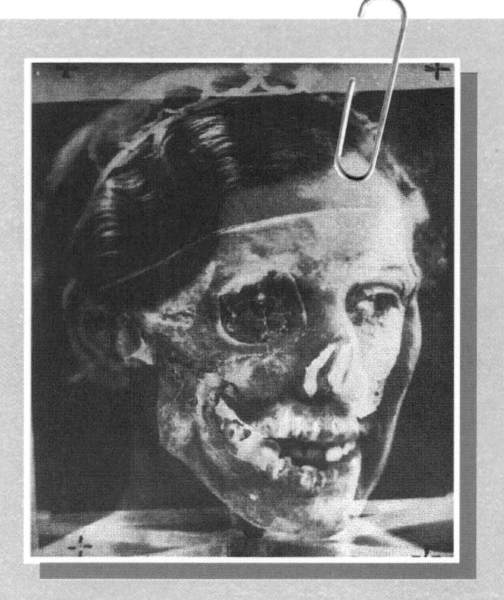

sign that the body was older than this, and the age was set at 35 to 45. Since the entire skeleton was available, the stature was set at 5 feet 3 inches (1.57 metres).

Although the bodies were already in an advanced stage of decomposition, it was relatively easy to estimate the time of death. One of the pieces of newspaper found with the bodies came from the Sunday Graphic of September 15, and the finding of some body fragments downstream from the bridge suggested that they had been carried there by a river flood on September 19.

The police therefore focused their inquiries on persons reported missing between these two dates. They learnt that the newspaper was a special edition distributed only in the Lancaster and Morecambe area of Lancashire. By chance, the Chief Constable of the county of Dumfriesshire had read of the disappearance of a young woman from Lancaster. Her name was Mary Jane Rogerson, a 20-year-old maid in the household of Indian-born Dr. Buck Ruxton (his original name was Bukhtyar Rustamji Ratanji Hakim). He had reported her disappearance, and that of his 34-year-old wife Isabella, to Lancaster police, and had subsequently given conflicting explanations of why they were missing. On October 12 he was formally charged with Mary Rogerson's murder.

There was circumstantial evidence in plenty: carpets in Ruxton's home were bloodstained, and others had been burnt; there were traces of human tissue and fat in the drains; and items of the clothing that had been wrapped round the body fragments were identified by Rogerson's mother and another woman. Finally, fingerprints and a palm print from body No. 1 were identified as Rogerson's. But was the other body that of Isabella Ruxton?

Professor Brash came up with a pioneering concept in forensic science – one that has since been employed in many cases. He obtained a good quality photograph of Isabella Ruxton, and photographed the skull of body No. 2 from the same angle. When the two photos were superimposed, they matched exactly. As a further piece of evidence, Brash made models of the two women's left feet in a flexible gelatin-glycerin mixture. That from body No. 1 fitted Mary Rogerson's left shoe, and that from body No 2. fitted Isabella Ruxton's.

Buck Ruxton stood trial for two murders, and was condemned to death. He was hanged on May 12, 1936.

The pieces of clothing in which some of the fragments of bodies were wrapped were identified by Mary Jane Rogerson's mother and another woman.

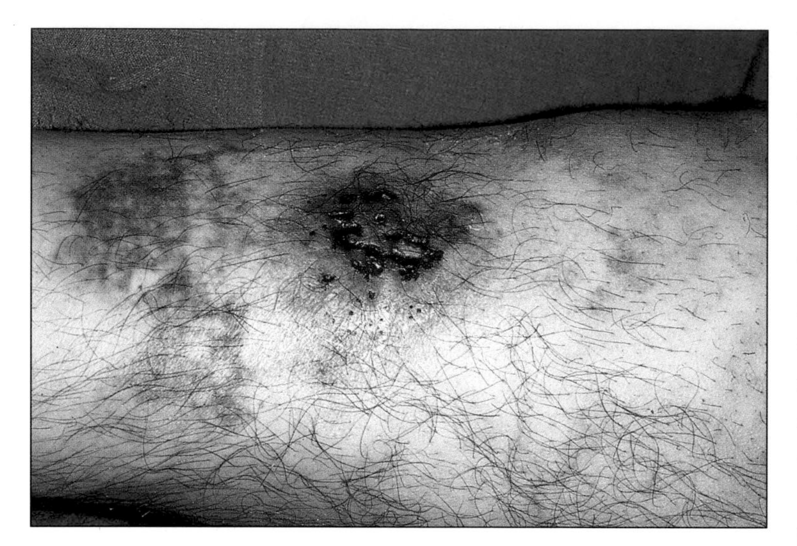

A human bite mark on a human leg. Residual saliva in this wound can be used to test for blood typing or DNA analysis.

bite out of a piece of cheese, and what was left was shown exactly to match his front teeth. A similar case occurred in England in 1984, when Arthur Hutchinson was tried for three murders, rape, and aggravated burglary. A wealth of evidence included bloodstains of a rare group, a print of his shoe, a palm print on a champagne bottle – and the two sets of bite marks he had left on a piece of cheese in a refrigerator.

Bite marks on flesh, however, present more of a problem. If there is no penetration, the underlying bruising may take up to 4 hours to develop in the living, and is clearly visible for up to 36 hours. Ultraviolet light can reveal bite marks even months after they have been inflicted. In a dead victim, they may take 12 to 24 hours to become visible. Usually, the only permanent evidence will be photographs of the injury – although it is possible to make a silicone rubber cast if the bite is deep enough. It is essential to take swab specimens from the site before casts are made, or the tissue is examined by the pathologist, as residual saliva can often be detected for blood typing or DNA analysis.

The odontologist must first assure himself that the bites are of human origin. Differences between human and animal bite marks are as follows:

- Human: U-shaped. Marks made by the canine teeth are not pronounced, and the indentations are broader and blunter in appearance than those of the following animals.
- Dog: A narrow squarish arch, with prominent pointed marks made by the canine teeth.
- Cat: A small rounded arch, with puncture marks from the canine teeth
- There are generally also scratch marks from the animal's claws.
- Rodent: Small bites, with long grooves caused by the incisors.

In non-sexual attacks the teeth are employed as a weapon: the bite is forceful and swift, there is no sucking of the flesh, and the general reddening caused by the rupture of tiny blood vessels is absent. Sometimes a finger, the tip of the nose or tongue, or an earlobe, may be bitten clean off.

Sexual attacks may combine a swift bite with the slower sucking popularly known as the "love bite". Here the teeth are used to grip during sucking, producing the characteristic reddening, either as a central patch or a peripheral ring, but often there are no visible teeth marks.

Odontologists have had to work hard to convince a jury that a photograph can establish identity with the dental characteristics of the accused. Two mid-20th century cases in Britain helped to establish the principle, and Dr. Keith Simpson was involved in both. In one of these, the battered body of Mrs. Margaret Gorringe was found behind a dance hall on New Year's Day 1948. There was a clear bite mark, made by two upper and four lower front teeth, on her right breast. Her husband was immediately suspected, and Simpson himself took a cast of his teeth. "Fortunately," wrote Simpson, "the suspect's teeth were so badly spaced and angled and curiously-shaped that there were a number of direct comparisons to make, and on every point these irregular features proved identical."

Reconstruction

The examination of skulls and teeth can provide vital evidence for the identification of bodies, and the determination of the cause of death, but there are many situations in which it is inconclusive. If, for instance, there is more than one body, and each is in scattered pieces, the skills of a specialist anatomist – a forensic anthropologist – will be needed. This is particularly important in such cases as aircraft disasters. On the other hand, a lone skull, or one associated with a skeleton that has not been identified, requires the dedicated art and imagination of a forensic sculptor, who can produce a likeness based solely on the bone structure and a detailed knowledge of facial types.

There are several claims as to who first made a successful facial reconstruction from a skull. In 1895 Swiss-born anatomist Wilhelm His acquired a skull believed to be that of Johann Sebastian Bach, and sculpted a face that was hailed as a good likeness. In 1916, an unknown skeleton was found in a cellar in Brooklyn, New York. A police anatomist mounted the skull, using a roll of newspaper for a neck. He fitted brown glass eyes, and covered the whole with flesh-coloured plasticine, which was then finished off by a sculptor. Several Italian residents immediately recognized it as that of Domenico la Rosa, who had vanished some time previously.

The Russian Mikhail Gerasimov was the most famous developer of this technique. In 1927, at the impressively young age of 20, he was put in charge of the department of archeology at Irkutsk museum. Prior to his appointment, he had spent two years measuring and dissecting the heads of corpses, to obtain reference data on how thick the flesh was on different parts of the skull, and the influence of the muscular structure. Using his findings, he began experimenting with the skulls in his care. In 1935, Gerasimov had his first successes, producing facial likenesses of people unknown to him, that closely resembled their photographs. In 1939, he was instrumental in securing the arrest of the murderer of a young boy. He finally saw his work crowned in 1950, when the USSR Academy of Sciences established its Laboratory for Plastic Reconstruction.

CRIME FILE:
John Wayne Gacy

A respected citizen in a Chicago suburb, he was really a homosexual serial killer. The jumbled remains of his young male victims presented an insuperable problem to forensic anthropologists, and many were never identified.

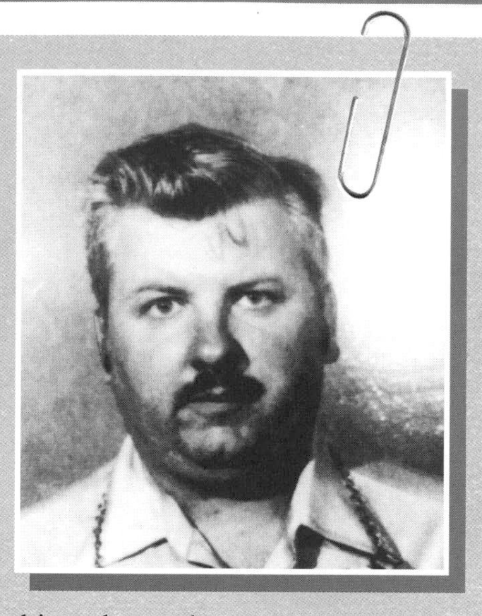

John Wayne Gacy – part-time children's entertainer, who was also known as "The Fat Man" – was a homosexual killer of 33 young men.

To the people of Des Plaines, John Wayne Gacy was a portly local builder who often took part in charity events dressed as a clown. He was also a leading light in the Junior Chamber of Commerce.

On December 11, 1978, 15-year-old Robert Piest disappeared. Local police discovered that he had been at Gacy's home; and a check on the builder's background revealed a less savoury picture. Gacy had a record of sexual offences stretching back more than a decade. In 1968 he had been sentenced to 10 years in Iowa State Reformatory at Anamosa for sodomy, but had been released on parole after 18 months. In 1971, only weeks after his release, he had been charged in Chicago with a similar offence, but the case was dismissed when the boy – another 15-year-old – failed to appear.

Armed with a warrant, the police returned to Gacy's house. Opening a trapdoor in the floor, they discovered a mass of human remains rotting in a sea of stinking black mud. They turned out be those of young males – most of them strangled. At police headquarters, Gacy confessed to killing 32 young men and boys over the past five years, 27 buried in and around his property, and five thrown into the nearby river. He had forgotten two: police eventually recovered 33 bodies at his home. One of four bodies recovered from the river proved to be Robert Piest's.

The Cook County Medical Examiner, Dr. Robert Stein, was faced with the problem of identification. The police had a long list of missing males but, because the killings had been homosexual, many

Police continue the grisly task of removing bodies from below the floor of Gacy's house in Des Plaines, December 1978.

By the end of January 1979, the bodies of only ten young men had been identified. Robert Piest is second from the right in the lower row.

parents were very reluctant to give any assistance. By the end of January 1979, only 10 had been identified, from dental records, X-rays and fingerprints, and after several frustrating months Dr. Clyde Snow – who was then working on the identification of bodies from the O' Hare aircraft crash – was called in. To assist him he brought Cook County radiologist John Fitzpatrick.

Snow's first task was to make sure that bones and tissues were matched up, and he then prepared a 35-point reference chart listing the characteristics of each skull. That each was from a male was soon confirmed, and examination of the teeth, cranial sutures and fusion of the epiphyses provided an approximation of age.

One missing person report was of 19-year-old David Talsma, who had broken his left arm as a boy. Snow found that one of the re-assembled skeletons had sustained such an injury. One of the skull sutures had closed early, and looked somewhat flatter than normal; hospital records from Kentucky that showed David had been treated for a minor skull fracture. The stature of the skeleton matched that of the missing youth. Finally, Snow found that the left arm was slightly longer than the right, and the shape of the shoulder blade socket suggested that he had been left-handed – as Talsma had been.

The work went slowly, and by the end of 1979 only five more bodies had been identified. Snow decided to call in Betty Ann Gatliff, a leading facial reconstruction expert from the headquarters of the Civil Aeromedical Institute in Oklahoma. She sculpted likenesses on nine still unidentified skulls – unfortunately, nobody came forward to say that they recognized any of the photographs that were published in newspapers. As Gatliff recalled:

> Two girls from different suburbs gave the same boy's name, and said he was their brother. But when it came to list the parents they said, "Oh no, I'm not going to tell you, because my mother just refused to talk about this." Well, we asked, who was his dentist? They said, "We don't have any idea." We said, could you ask your mother? "Oh no, she won't even talk about it."

Several years later, a local newspaper reporter managed to identify one of the faces that Gatliff had reconstructed. The rest were simply buried as "John Does".

CASE CLOSED

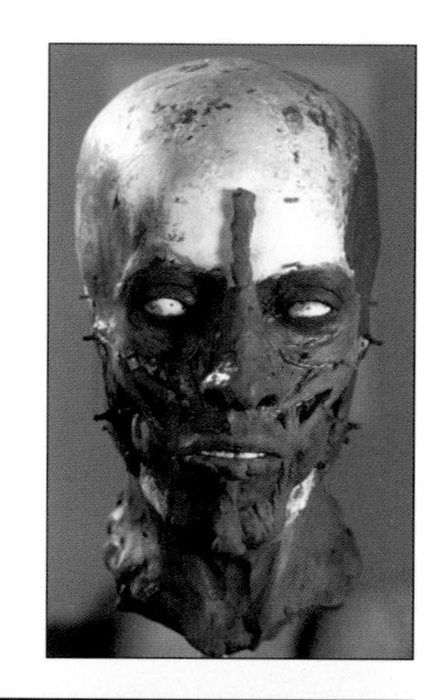

Facial reconstruction (right and far right), made by Richard Neave, Director of Manchester University's Unit of Art in Medicine, in 1989 to assist in the "little Miss Nobody Case". Photographs of the finished reconstruction were distributed to the press and television, and the victim was soon identified as Karen Price.

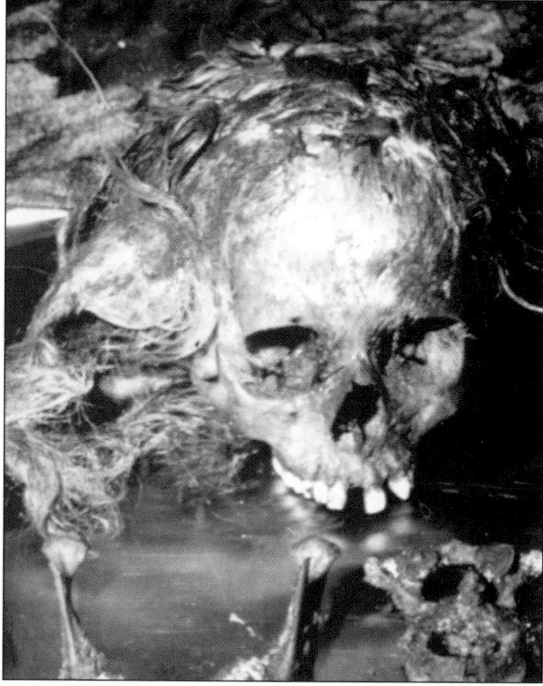

In the Karen Price case, Richard Neave started his reconstruction work with the skull of the uncovered remains (above).

For many years the Russians were the foremost in facial reconstruction.

At present, Britain's Richard Neave, of Manchester University's medical faculty, is one of the world's leading figures in the technique. In 1989, building workers in Cardiff, south Wales, discovered a skeleton wrapped in a carpet. Pathologists, an odontologist and a forensic entomologist established that it was that of a young girl, aged 15, who had been buried at some time between 1981 and 1984. But what the police wanted was the face of the victim, whom they named "little Miss Nobody".

Richard Neave spent two days over his reconstruction. At the same time, a laser-computer analysis (see Fact File opposite) was made by Dr. Peter Vanezis at the London Hospital, but in the event it was not needed. Photographs of

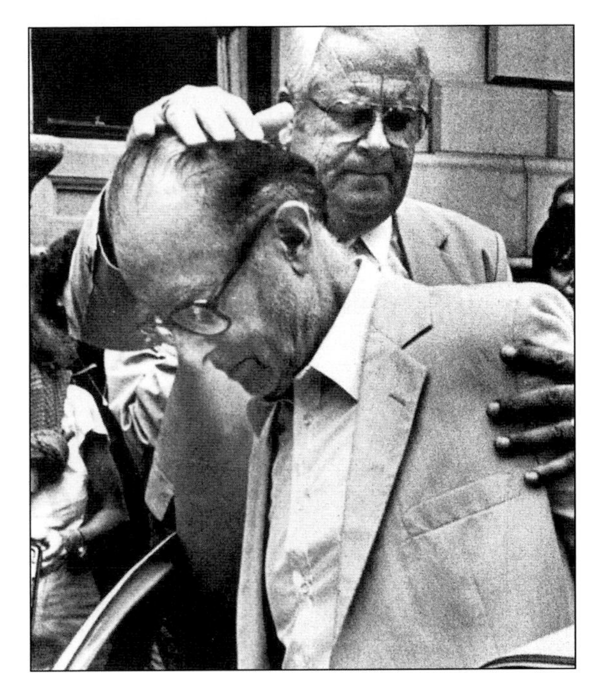

Neave's work were distributed to the press and television, and just two days later a social worker reported that they resembled a Karen Price. Price's dental records were found, and confirmed the identification. Finally, DNA extracted from the bones of the victim was compared with that from the blood of Price's parents, and the identification was complete.

The police soon uncovered Price's unhappy history. A runaway, she had taken to prostitution and, when she had refused to pose for pornographic photos, her pimp and a doorman from a local bar had killed her in a rage. Both were found guilty in February 1991.

A rather different case is that of John List, who murdered his wife, mother and three children in New Jersey in 1971 – and then disappeared. He remained on the FBI list of wanted men, and in 1989 sculptor Frank Bender produced a head based on a photograph of List, but showing how he might look, aged by 17 years. When it was shown on the television programme *America's Most Wanted* it produced hundreds of telephone calls, many naming a man called Robert Clark, living in Richmond, Virginia. He might have changed his name, but he could not change his fingerprints; he was arrested, charged with four murders, and found guilty.

John List, discovered after more than 17 years.

The basic technique for facial reconstruction is very simple. Casts of the skull are made in a flexible but firm plastic, firstly as it is received, to preserve any adhering features, then again after it has been completely cleaned. The eye sockets are filled with balls of polystyrene then, at important anatomical points, small holes are made in the surface of the cast, and fitted with small wooden pegs – similar in thickness to tooth picks – to the expected depth of the flesh at these points.

The muscles and other features are then built up with modelling clay to the height of the pegs. The shape of the nose and ears is difficult to estimate, as the skull provides little evidence, and it is here that the experience of the sculptor comes into play. The cheeks and temples are filled out, and the thin flesh over the scalp is laid on in strips.

When the clay has been made smooth, artificial hair is used to add eyebrows. A wig, or a head of hair, may complete the reconstruction.

Since the late 1980s, a very different technique has become available. The skull is set on a turntable, and a laser beam is reflected from it as it turns. The information is fed to a computer, where it is compared with data obtained from the head of a living person with similar skull measurements, and assembled into a likeness.

CRIME FILE:

Alferd Packer

The "Colorado cannibal" freely admitted that he had kept himself alive through a bitter winter by eating the bodies of his companions. But had he murdered them, or successfully defended himself against the attack of the real killer?

Alferd Packer, the Colorado Cannibal, who survived being holed up in the Colorado mountains during the winter of 1874 by eating his five fellow prospectors.

In the bitterly cold winter of 1874, a party of six prospectors was holed up deep in the Colorado mountains. When the snows melted in the spring, only one emerged, looking suspiciously well fed. His name was Alferd Packer.

When he was arrested, Packer did not deny that he had eaten his five companions. He claimed that they had either died of natural causes or had been killed by the fifth, Shannon Bell, and that he himself had saved his life only by killing Bell in self-defence. He was sentenced to death for murder in 1883, although a subsequent re-trial commuted the sentence to 15 years in prison.

Packer became, over the years, something of a local hero, and in 1989 James Starrs, professor of law and forensics at George Washington University in Washington DC put together a team of archeologists, anthropologists and pathologists to search for the remains of the victims. They found enough to account for five, but unfortunately none was complete, and their contemporary descriptions were insufficient for identification. Nevertheless, Professor Starrs was convinced that "Packer was guilty as sin, and his sins were all mortal ones".

Marks on the bones made it clear that at least four of the victims had been bludgeoned to death with a hatchet, then carefully defleshed with a skinning knife. Anthropologist Walter Birkby, of Arizona State museum, spent nearly a month studying the bones. His verdict was more cautious than that of Professor Starr. "We'll never know who did it based on any physical evidence," he said. "We're never going to know."

CASE CLOSED

Scattered Remains

One of the most difficult problems facing forensic investigators is the sorting-out and identification of the confused remains of more than one victim, which may be scattered over a wide area, and of which some parts may never be recovered. Most often this is not the result of a criminal act, but due to such disasters as aircraft or train crashes and fires. Sometimes, however, the remains have been distributed by a murderer, and further scattered by animals, or they form the charnel-house of a psychopathic killer.

Anatomists have a detailed knowledge of the proportions of the human body, and are frequently able to match up limbs with their related trunks, or vertebrae attached to a skull with the rest of the spinal column. Tissue analysis and bone samples can provide further evidence, but when there are many victims, as in an airport crash, the task of identification is a formidable one.

A new speciality, known as forensic anthropology, has developed in the United States in recent years. Foremost among its practitioners is Dr. Clyde Snow, who spent many years at the Civil Aeromedical Institute of the Federal Aviation Authority (the FAA), where he studied what happened to people in aircraft crashes. He prefers to call his work "osteobiography", because "there is a brief but very useful and informative biography of an individual contained within the skeleton, if you know how to read it".

Snow resigned from the FAA in 1979 to concentrate on forensic matters. He was involved in many investigations, including the examination of the exhumed body of Dr. Joseph Mengele, the Nazi concentration camp experimenter; the identification of the remains of the "disappeared" in Argentina, Bolivia and Guatemala; the examination of the 33 victims of serial killer John Wayne Gacy; and the excavation of mass graves in the former Yugoslavia. However, his role in identifying the fragments of the 273 victims of the O' Hare aircraft crash of 1979 was probably the biggest challenge of his career.

The huge explosion and fire had broken the bodies into pieces, and consumed most of their clothing and documents that could give a clue to identity. A fireman on the scene said: "We didn't see a single body intact .... just trunks, hands, arms, heads, and parts of legs. But we couldn't tell whether they were male or female, whether they were an adult or a child, because they were all charred."

Many victims were fairly quickly assembled and identified by the teams of pathologists, dentists, fingerprint and other experts working on the investigation, but Snow was left with nearly 50 that defied identification.

Lowell Levine, consultant odontologist to New York City Examiner's office, reported that the operation involved "thousands and thousands of little bits of paper." Snow suggested the use of a computer and, with the help of a programmer from American Airlines, entered everything that was known about the still-unidentified passengers, together with the anthropological and dental details of each victim. He worked in shifts with radiologist John Fitzpatrick, and in five weeks they identified 20. But with no further information, 29 bodies could not be fully identified.

Despite the many successes that have been recorded by investigators, the science of forensic anthropology remains an uncertain – although extremely valuable – technique. Skulls and bones can tell their story, but it is sometimes an incomplete one.

Breath of Life

DEATH BY OXYGEN DEPRIVATION is known medically as asphyxiation. In life, the lungs transfer oxygen to the blood, where it is carried by hemoglobin to organs and tissues throughout the body and, most importantly, to the brain. Without oxygen, the body will quickly die: if the lungs and heart cease working, death is imminent. As the British physiologist J.B S. Haldane pointed out in 1930, oxygen deprivation "not only stopped the machine, but wrecked the machinery". Modern resuscitation techniques are directed at getting the machine to work again; even if the heart has stopped beating, the brain can remain alive without oxygen for up to ten minutes – although gradual neurological deterioration will occur during this time.

There are more than 25,000 species of diatoms, microscopic algae, usually single-celled, which occur in both fresh and salt water. Their cell walls contain silica, clearly visible in these specimens of Campylodiscus hibernicus.

Asphyxiation occurs when the lungs are unable to take in air. It can be produced in many ways. Intense pressure on the thorax, which prevents the movement of the lungs, is the cause of death in crowd disasters, or when an excavation collapses on the victim. Blockage of the windpipe, or choking, can happen accidentally – by swallowing food too fast, for instance, or by inhalation of water (drowning). Blockage can also be induced deliberately, either by suffocation or strangulation, either manually or with a ligature of some kind – a cord, a rope, a tie or a scarf: anything that can be twisted about the throat.

Strangulation may cause death in two other ways. Pressure on the windpipe can directly cut off the blood supply to the brain. Alternatively, it can overstimulate the vagus nerve, which detects variations in the blood pressure in the carotid artery. The brain responds to the increased pressure by stopping the beating of the heart.

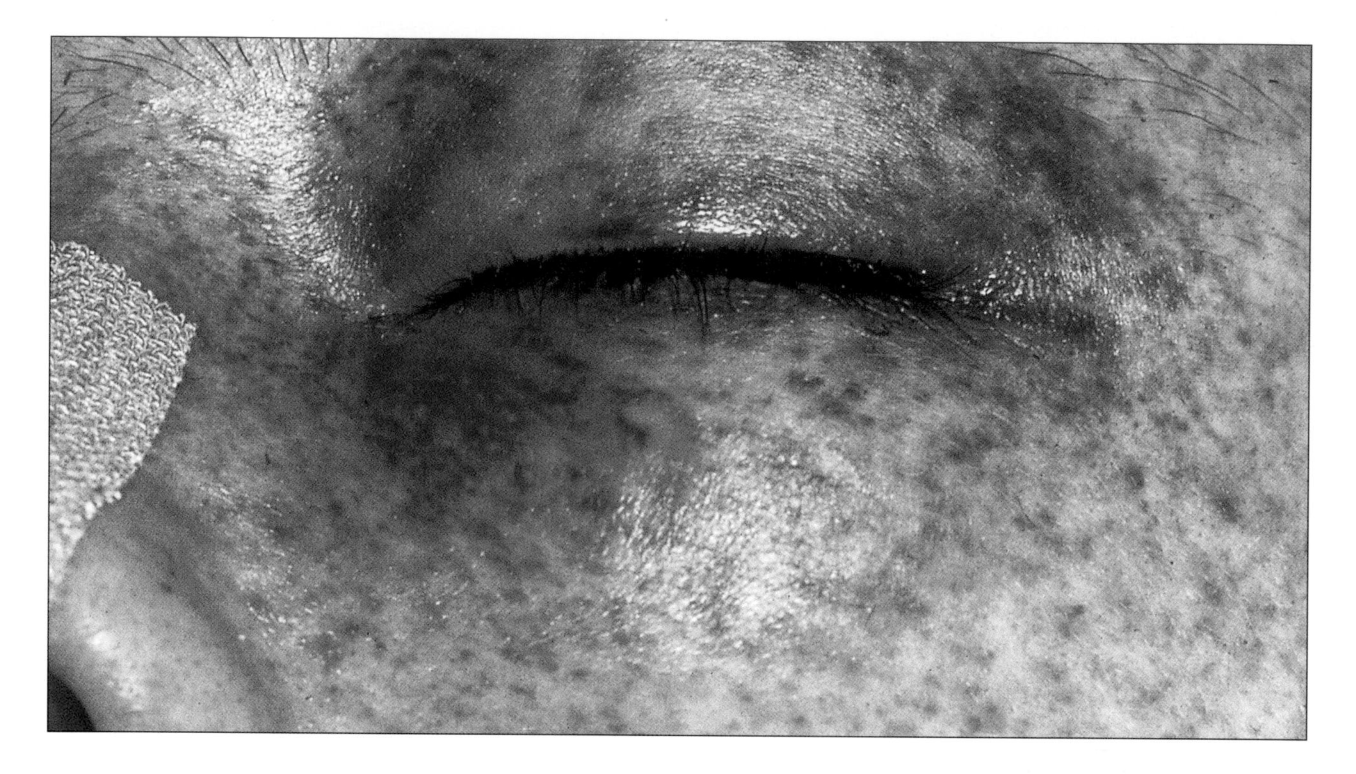

A common sign of asphyxiation is the formation of petechiae – tiny ruptured blood vessels – on the eyelids and the skin of the face.

There are often obvious signs of asphyxia. The face becomes swollen, due to raised pressure in the veins. The skin, particularly that of the head and neck, becomes visibly blue – a condition known as cyanosis – being coloured by oxygen-free hemoglobin, such as is seen in the veins. Tiny ruptured blood vessels – petechiae – can be found, particularly in the whites of the eyes, the outer eyelids, the skin of the face, the lips, and behind the ears. Surprisingly, the body temperature tends to rise for an hour or two after death by asphyxia.

SUFFOCATION AND STRANGULATION

Smothering with a pillow or some other soft material – or even the hand – leaves very little telltale signs for the forensic investigator; the same is true of suffocation inside a plastic bag. This is because the victim is almost always an old person in bed, or a young infant, who is unable to offer resistance. Cyanosis and petechiae do not usually develop unless there is a struggle; then there may be minor bruises, or abrasions inside the mouth where the lips have been pressed against the teeth. However, petechiae may develop if the body is lying face-down.

Homicidal strangulation may be carried out manually or with a ligature. In both cases it leaves clear signs. Manual strangulation usually indicates that the assailant is bigger and stronger than the victim. For this reason it is generally committed by a man against a woman, often during, or subsequent to, a rape.

CRIME FILE:
Catherine Fried

She believed that she had committed the "perfect murder". Only the testimony of a confessed killer, and the subsequent re-examination of medical records, led to the uncovering of vital evidence.

Catherine Fried believed she had asphyxiated her husband Paul without trace, until later examination of records revealed petechiae and other evidence.

The body of 61-year-old Paul Fried, a prominent Philadelphia gynecologist, was discovered, lying face-down on the floor of his bedroom, by his wife Catherine on July 23, 1976. Although they had been married only a year, the couple were living apart, but still saw one another regularly. Catherine told the police that her husband was addicted to alcohol and barbiturates: worried because her telephone calls went unanswered, she had come to the house and discovered the body. Fried's nose had bled, and a pillow stained with some of the blood lay over his head.

On the bedside table the police discovered a scrawled note that they took to be evidence of suicide. The assistant medical examiner made a brief examination, and signed the death certificate "drug overdose". Catherine Fried had the body embalmed, and made arrangements for cremation. But Fried's daughters from a previous marriage were dissatisfied, and did not believe their father had committed suicide. They halted the cremation, and asked the retired former chief medical examiner of New York City, Dr. Milton Helpern, to carry out a second autopsy.

Helpern wrote a 15-page report, in which he waxed lyrical on the human body as a "museum" of the ageing process, but none of the pathological signs he discovered was imminently life-threatening. He noted pinpoint petechiae in Fried's eyes, and minor bruises on his neck – nevertheless he gave his opinion that death was due to natural causes.

Some time later, a man named Jerald Sklar, who had worked with Catherine Fried, approached the FBI. He asked for witness protection, and told them that he and a friend had murdered two men, and that Fried had paid him money to secure her husband's death. When he refused, she had herself suffocated her husband with the pillow.

The Philadelphia DA was unsure about prosecuting for murder on this evidence alone. He asked Dr. Helpern's successor, Dr. Michael Baden, for his opinion. As Fried's body had been cremated, Baden had only the records and photographs to review. He was struck by Helpern's observation of the petechiae and minor bruises. The toxicology report, which had returned from the lab only after the death certificate was signed, had been ignored – even though it showed the presence of very little barbiturate, and no alcohol. Suicide was ruled out.

At Fried's trial, Baden gave his opinion that her husband's death was "consistent with suffocation", and she was found guilty. She appealed on grounds of insufficient evidence, and a second trial was heard. Again she was found guilty.

CASE CLOSED

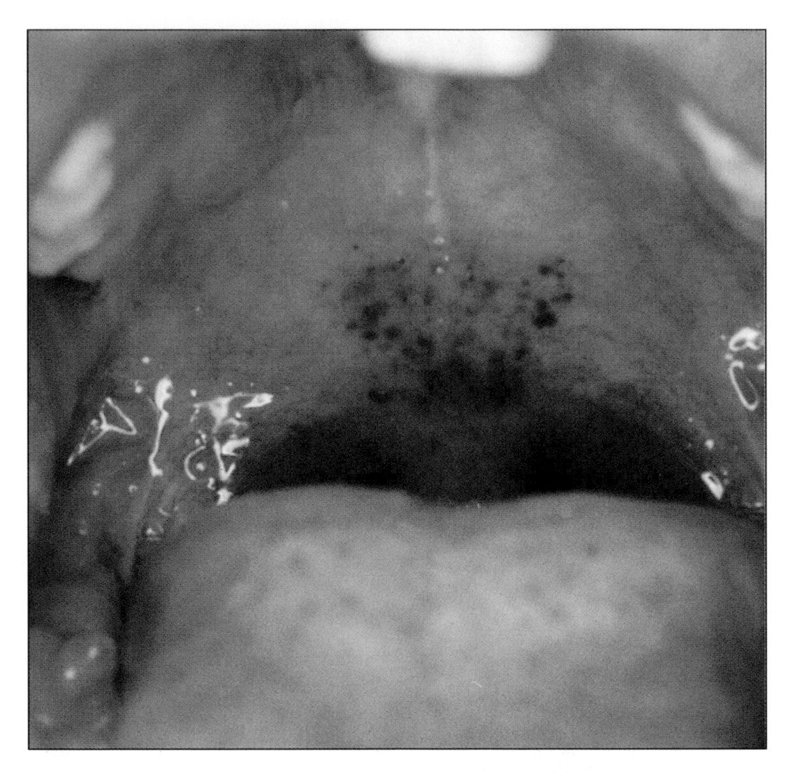

Petechiae, sure signs of asphyxiation, may also be found inside the mouth.

One or both hands may be used to strangle the victim, from in front or from behind. The external signs are bruises and abrasions on the neck, frequently following the jawline. The bruises from the fingers are generally disc-shaped, a little smaller than the pads of the fingertips, but the thumb will cause a slightly larger mark. However, the grip – and even the hands – may change during the attack, resulting in larger, irregular bruising.

If the strangler has kept the pressure up for some time, the characteristic congestion and petechiae will be seen in the face of the victim. If, on the other hand death takes place within a few seconds, they do not appear. This is particularly true if the actual cause of death is sudden cardiac arrest.

On internal examination at the autopsy, the larynx may be found damaged, with the upper horns of the thyroid cartilage fractured on one or both sides. This is less common in young people, whose cartilage is pliable, and not brittle. The horns of the small hyoid bone below the jaw may also be fractured, and there are other injuries that may occur if blows have been aimed at the throat. Hemorrhage will also occur in the neck muscles.

Rope, wire, string, electric or telephone cable can act as the ligature used to strangle a person, as can a strip of cloth, or an item of clothing – a scarf, necktie, stockings, even a bra. If the ligature is knotted, there may be certain peculiarities about the knot that can sometimes provide a clue to the murderer. In 1962, Boston police were first alerted to the fact that a serial killer – the "Boston Strangler" (later identified as Albert

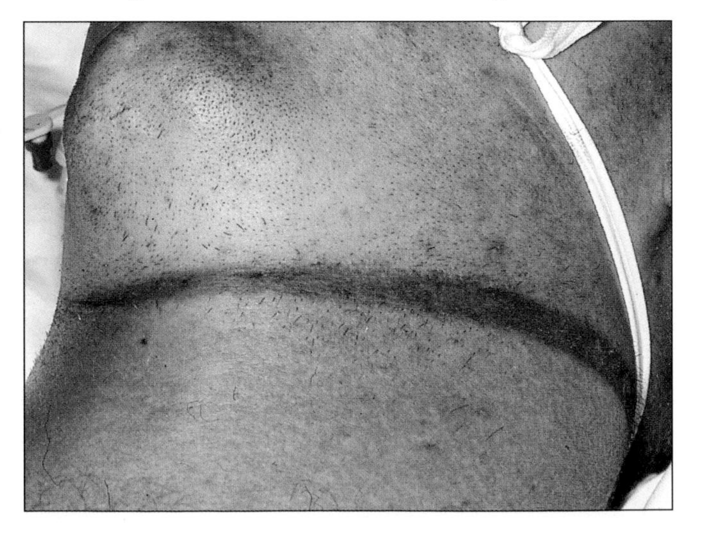

In strangulation, the ligature used – cord (as here), rope, wire, or even a strip of cloth – will leave a distinctive mark in the flesh. If the ligature is missing, the examining pathologist will use a strip of Scotch tape to lift any remaining fibres from the injury.

The "Boston Strangler", Albert DeSalvo, was finally apprehended in 1964. On many of his victims, he had left a "signature" bow loosely tied in the ends of the ligature he had used to strangle them.

DeSalvo) – was at work by the way he tied the ends of the ligature into a floppy bow after his victim was dead.

Congestion, cyanosis and petechiae will develop if the pressure is maintained for more than 15 seconds. Since the ligature must be tight enough to constrict the neck, it will make a distinctive mark in the flesh. This evidence is particularly important when the assailant does not leave it in place, as the mark can reveal the nature of the ligature and its width. If it is still present, the examining officer must take care to preserve the knot when cutting it.

The position of the ligature mark is generally almost horizontal, above the larynx, and goes around the neck. If wire or a thin cord was used, the mark will be clear-cut and deep, with raised edges. A strip of soft fabric is likely to leave a relatively faint and undefined mark; a broad cloth, such as a scarf or towel, can leave one or more narrow marks, where the stretched fabric has been drawn into tight narrow bands. If the killer has attacked from behind, perhaps holding a cord stretched between two hands, the ligature mark will be only across the front of the throat. There are likely to be scratches on the neck, where the victim has struggled desperately to loosen the ligature.

Internally, the damage is similar to that from manual strangulation, but is generally less severe.

CRIME FILE:

Harold Loughans

Was it possible that he could have strangled a defenceless woman with his mutilated hand? One forensic expert said yes, another disagreed. The truth only emerged 20 years later.

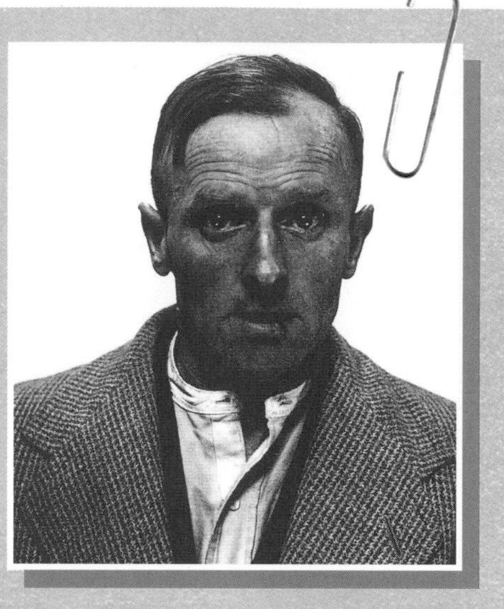

Harold Loughans, photographed by police at the time of his arrest for the murder of pub licensee Rose Robinson.

The body of Rose Robinson, licensee of the John Barleycorn pub in Portsmouth, southern England, was discovered in her bedroom on the morning of November 29, 1943. She had been strangled. There were signs that she had tried to open the window to call for help, and £450 was missing from the room. Dr. Keith Simpson examined the body:

> I thought she had probably been strangled as she lay on the floor, with her murderer either kneeling or sitting astride her. The fingermarks told a clear story: a deep bruise on the right of the voicebox, presumably made by the thumb, and three lighter bruises in a line on the other side. Right-handed, four inches across…

A month later, a petty thief named Harold Loughans was arrested in London; he confessed to the murder. Trace evidence linked him with the crime scene, but there was one apparent flaw in the case against him – he lacked the first two joints of the fingers of his right hand.

Loughans was tried in March 1944, and withdrew his confession. Four witnesses swore they had seen him in London on the night in question. The jury failed to agree on a verdict, and the case was heard again two weeks later. This time, the defence held another trump card: the most respected pathologist of the day, Sir Bernard Spilsbury testified that he had visited Loughans in prison, and asked him to hold his hand "with all the strength he had". "I do not believe," said Spilsbury, "he could strangle anyone with that hand." Loughans was acquitted.

In 1963, the prosecuting counsel in the case, J.D. Casswell, KC, published his autobiography, in which he implied that Loughans had been lucky to be acquitted, and extracts appeared in *The People* newspaper. Loughans, who had just been released from prison for other offences, brought a civil action against the newspaper for libel.

Throughout the case, prosecution and defence argued whether or not Loughans had sufficient strength in his right hand. Simpson produced his notes and drawings from 20 years before, and said that, once Loughans had Rose Robinson on the floor, he had only to put the weight of his body behind the hand to strangle her. The libel action was dismissed – in effect finding Loughans guilty of a crime for which he had already been acquitted, and for which he could not be tried again. Two months later, he walked into the offices of *The People* and, with his right hand, wrote: "I want to say I done that job. I did kill the woman in the public house at Portsmouth."

The maimed hand of Harold Loughans, which led Sir Bernard Spilsbury to assert that he was incapable of having strangled Rose Robinson.

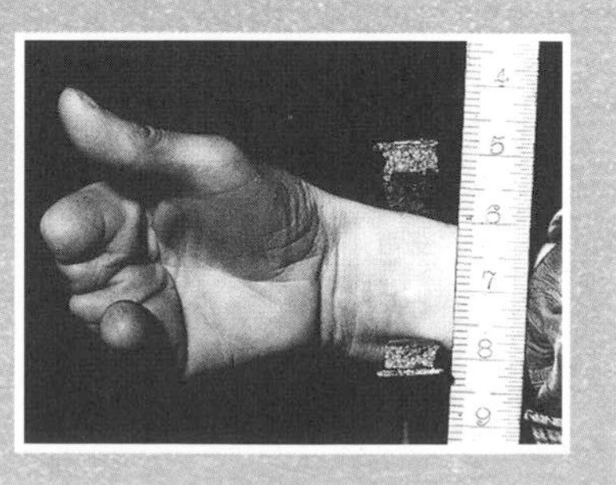

DROWNING

Death by drowning – a fate that one in every 40,000 of the population can expect to suffer – may be accidental, suicidal or homicidal. Accident is the most likely cause, and 80 percent of accidental cases involve men; homicidal drowning is relatively rare, and here the majority of incidents are tragic cases of infanticide. But when a body is "found drowned", a coroner's inquest is almost always required, and autopsy will be necessary to determine the exact cause of death. This is because there have been a number of attempts to disguise murder, by some other means, as accidental drowning.

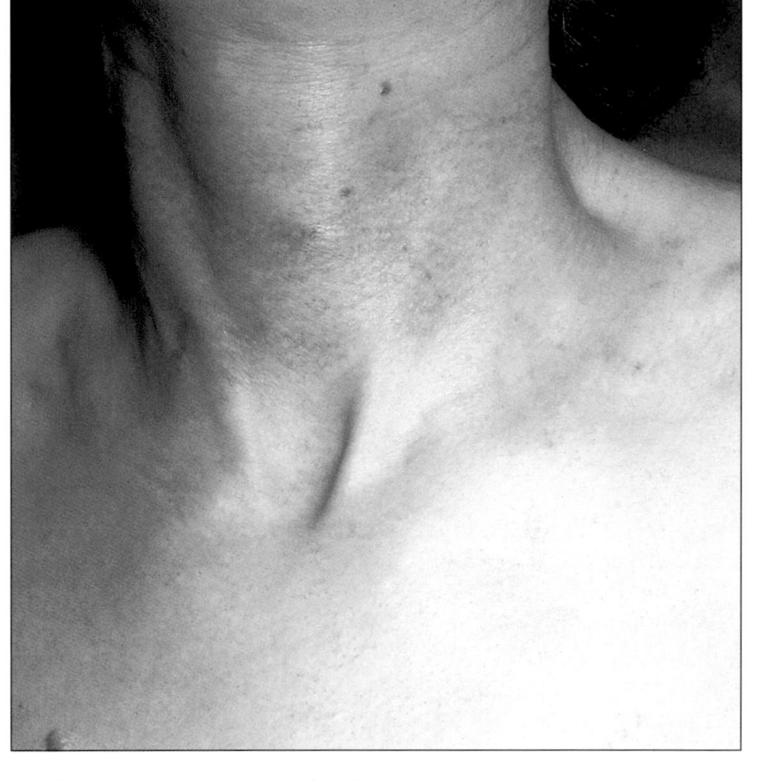

In a case of manual strangulation, the neck will be marked with bruises from the fingers. However, if the grip is shifted during the attack, the bruising will be larger and irregular.

Drowning can occur in a river, a lake or a canal, in the sea, in a household or public bath, even in a puddle – many cases have been recorded of drunken people falling forward into pools of rainwater only an inch or two (a few centimetres) deep, and drowning. Moreover, one can drown in other liquids than water: a vat of beer, paint, or some chemical liquid in a factory, or even in one's own vomit. One of the most dangerous fluids of all is a thick mixture of sand and sea water, suddenly inhaled when one is knocked down in the shallows by a large wave, or struck in the belly by a surf board.

Death by drowning is not always the long struggle for life (with "all one's life passing before one") that is generally supposed. In as much as 15 percent of cases, death occurs within seconds. Pathologists describe this as "dry drowning", because little or no liquid may be found in the lungs. There appear to be two different mechanisms at work here. The shock of entering the water may cause instant heart arrest. Alternatively, water entering the nose can cause a spasm in the larynx; this is to prevent water entering the lungs, but as a result they are immediately deprived of oxygen, and unconsciousness can occur almost at once, rapidly followed by death. The influence of alcohol or drugs can be critical in drowning of this kind: drunken sailors who fall into the harbour on the way back to their ships are a common example.

"Wet drowning", when the water is inhaled, also deprives the lungs of oxygen, but here again other factors intervene. If the drowning occurs in fresh water, large quantities are rapidly absorbed into the bloodstream, whose volume may increase by as much as 50 percent within a minute. This places a great strain on the heart, which will soon fail. On the other hand, sea water has a salt concentration higher than that in the blood, causing water from the tissue to move into the blood vessels of the lungs, producing pulmonary edema. As this places no consequent strain on the heart, it is probably the reason why those drowning at sea can struggle to survive for a much longer time than in fresh water. The temperature of the water is also important. Cold water can cause death by hypothermia; the shock of sudden immersion may also cause immediate heart failure.

Finally, pathologists distinguish between primary drowning, in which death occurs too quickly to make resuscitation possible, and secondary drowning. In the latter, a resuscitated victim may survive for as much as several weeks before finally dying of edema or a secondary infection.

The most characteristic sign of drowning is the production of a frothy liquid in the lungs, which is forced out through the larynx and generally appears around the mouth and nostrils. This is a mixture of water and mucus whipped into a foam by violent attempts to breathe. It is often tinged with blood from the rupture of small blood vessels in the lungs. This froth begins to disperse quite rapidly, however, and is not found in bodies that have been in the water any considerable time.

Occasionally, instantaneous rigor (see "Gathering the Evidence") occurs at the moment of death, and the body may be found still clutching at grass or weeds. If the hands are holding fragments of clothing, or the hair of another person, suspicions of homicide must inevitably arise, particularly if there are signs of a struggle on dry land. Yet in one instance, the body of a man was found floating in a canal with his throat cut, and a trail of blood leading from a bridge back to an empty house, where police discovered an open razor on the floor. It looked like murder, but the only prints on the razor were the man's own. It appeared that, having cut his own throat, he had then managed to stagger to the canal and thrown himself in to drown.

When a body is recovered after a period of immersion, there are no external signs to specify whether it was alive or dead at the time it entered the water. The skin will be

FACT FILE

The earliest treatise on forensic medicine is the ancient Chinese book *Hsi Yuan Lu* (*The Washing Away of Wrongs*). It contains much unscientific advice, but it also includes details on how to distinguish between strangulation – characterized by pressure marks on the throat and damaged cartilage in the neck – and drowning, in which water is found in the lungs. It is remarkable, therefore, that it was not until 1890 that western doctors conclusively established that water entering the lungs caused death by drowning.

CRIME FILE:

George Joseph Smith

He became known as the "Brides in the Bath" murderer. How he had drowned three of his wives in order to cash in their life-insurance policies was demonstrated in a dramatic – and near fatal – way at his trial in court.

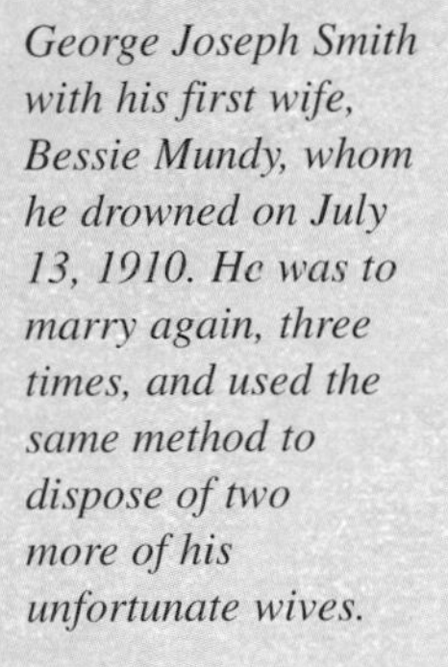

George Joseph Smith with his first wife, Bessie Mundy, whom he drowned on July 13, 1910. He was to marry again, three times, and used the same method to dispose of two more of his unfortunate wives.

George Joseph Smith was an English confidence man who made a living by persuading lonely spinsters to marry him, and then making off with their money. But when he married Bessie Mundy in 1910 he found that her inheritance was in trust, and so had to persuade her to make a will in his favour. On July 13, 1912, Bessie Mundy was found on her back in the bath, her head under water, with a piece of soap clutched in her right hand. The coroner's jury decided that she had had an epileptic fit in the bath, and had drowned.

On November 4, 1913, Smith married Alice Burnham, and insured her life for £500; on 12 December she, too, died in her bath. His next wife, Alice Reavil, was luckier; Smith married her on September 17, 1914, and left her £90 poorer a few days later. Finally, on December 17, 1914, he married Margaret Lofty, insuring her life for £700 – and by the evening of the following day her lifeless body lay in the bath.

A report in the *News of the World* on January 1, 1915, told of the "Bride's Tragic Fate on Day after Wedding". Alice Burnham's father read it – and so did the landlady of the seaside lodgings in which she had died. Both were struck by the similarities of the two cases, and reported their concern to the police. Smith was detained for questioning on February 1, and on the same day Dr. Bernard Spilsbury (as he then was) exhumed and examined the body of Margaret Lofty. Shortly after, he examined the other two bodies.

There was no sign of violence on any of the bodies, and no doubt that they had drowned. But how? Was this a case of serial murder or not? The women had been found on their backs, their heads below the sloping end of the bath, and their legs straight out of the water at the other, vertical, end. Spilsbury pointed out that if the dead women had died of a fit, their heads would have been forced up the sloping end – out of the water – by muscular spasm. On the other hand, there was no sign of a struggle, which might have been expected if Smith had forced them to drown.

At Smith's trial, Spilsbury demonstrated his explanation with dramatic results. A nurse dressed in a bathing costume got into one of the baths, and Detective Inspector Arthur Neill grabbed her feet and pulled her under the water. To his horror, the rush of water into her nose and mouth rendered her immediately unconscious, and she had to be revived by artificial respiration – but Smith's guilt was established. He was hanged in August 1915.

"goosey", and the feet and hands sodden and wrinkled – the so-called "washerwoman's hands". After about two weeks the skin loosens from the hands and feet, and in three to four weeks the entire skin may come away from the hands like a glove. In the so-called "Hand in Glove" case, a section of skin like this was found in a creek near Wagga Wagga, New South Wales, Australia, in December 1933. Experts were able to get a fingerprint from it, and establish the identity of the victim, without the rest of the body. He was an itinerant named Percy Smith, and his identification soon led to his murderer, Edward Morey.

A submerged body tends to be suspended in the water face-down, with the legs and arms also hanging down. Any hypostasis, therefore, usually develops in the face, the upper part of the trunk, hands and lower arms, calves and feet. The tumbling of waves, however, will affect the development of the hypostasis, which may not be very apparent.

The decomposition of a body in water proceeds more slowly than on land, and a water temperature of around 40°F (5°C) can retard putrefaction by several weeks. As decomposition continues, a considerable volume of gas is produced in the viscera and, provided the body is not caught among rocks or entangled in weeds, it will tend to float to the surface in the course of some two weeks. Identification of drowned individuals is sometimes very difficult, as the decomposition processes that follow death can cause the body – particularly the face – to become distended beyond recognition, and the skin can be coloured almost black by decaying blood.

One remarkable effect of prolonged immersion, or of burial in wet surroundings, is the occasional formation of adipocere (see "Gathering the Evidence"). This is due to chemical action on the body fat that changes it into a substance similar to soap. Adipocere is a greyish, waxy substance that retains the shape of the body – although the face is not often recognizable. It usually takes months to develop – although cases are known in which it was formed in just three or four weeks – and it can persist for years or even centuries.

Establishing whether death occurred before the body entered the water, or whether it was caused by drowning, is one of the most difficult things to decide at autopsy. In wet drowning the lungs will be waterlogged and spongy to the touch, but this does not happen with dry drowning. Many different methods have been suggested for the chemical examination of the blood in the two halves of the heart – the left side will be more affected than the right by the effects of fresh or salt water on the blood – but none of these has proved reliable.

The diatom test is potentially the most valuable means of establishing whether a case is one of drowning. Diatoms are microscopic organisms, which occur both in sea water and in unpolluted fresh water; there are at least 25,000 species, many distinguishable from one another by the shape of their acid-resistant silica shells. In a drowning person these diatoms reach the lungs and are absorbed into the blood, and while the heart continues to beat they are distributed round the body, lodging in organs such as the kidneys, brain and even the bone marrow.

If the body is dead when it enters the water, diatoms may enter the lungs, but cannot be distributed by the circulation. Carefully cut sections of the long bones of the body are extracted with nitric acid – making sure that these do not become contaminated with the water from which the bone was taken. The diatom content of the bone marrow is then compared with that of the water. In one case scientists, by identifying the characteristic diatom species, were able to show that a body washed up on the Belgian coast had entered the water from a yacht off the Isle of Wight, in the English Channel.

A micrograph showing many distinctive forms of diatom. When a drowning person swallows water, diatoms can enter the bloodstream. This is potentially the most reliable means of determining whether the person was alive on entering the water. Diatoms discovered in the body of the drowned magnate Robert Maxwell established that he had committed suicide, and was not the victim of foul play.

It has yet to be conclusively established that diatoms could not enter the bloodstream during everyday life. Nor do we know how long they will remain in the bone marrow. There are fierce arguments among pathologists on both sides, and the "diatomic war" continues. Fortunately, homicide by drowning is a rare event. There are usually signs of a struggle on dry land, injuries to the body that are not due to any damage it may have sustained while in the water, or evidence of drugs having been administered. Water may wash away our sins, but it cannot hide the sins of others.

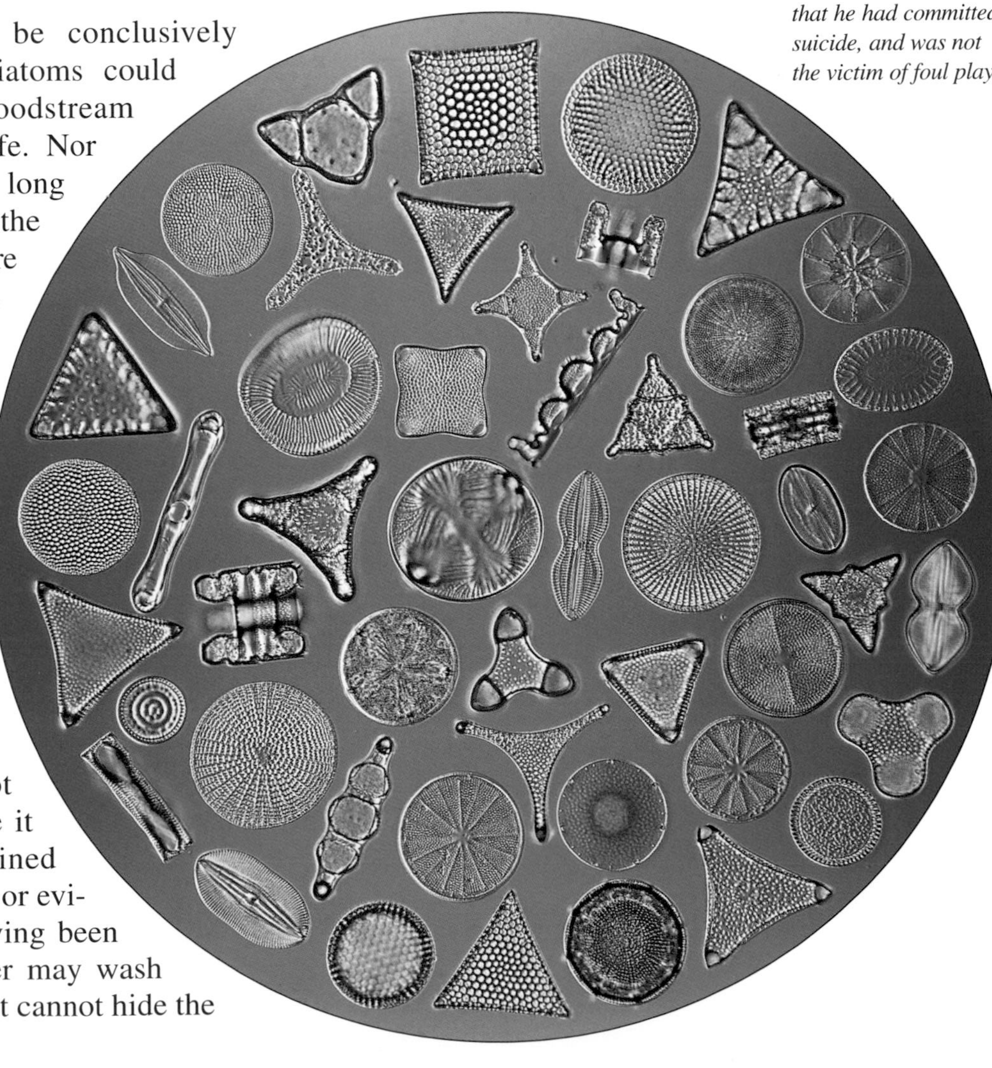

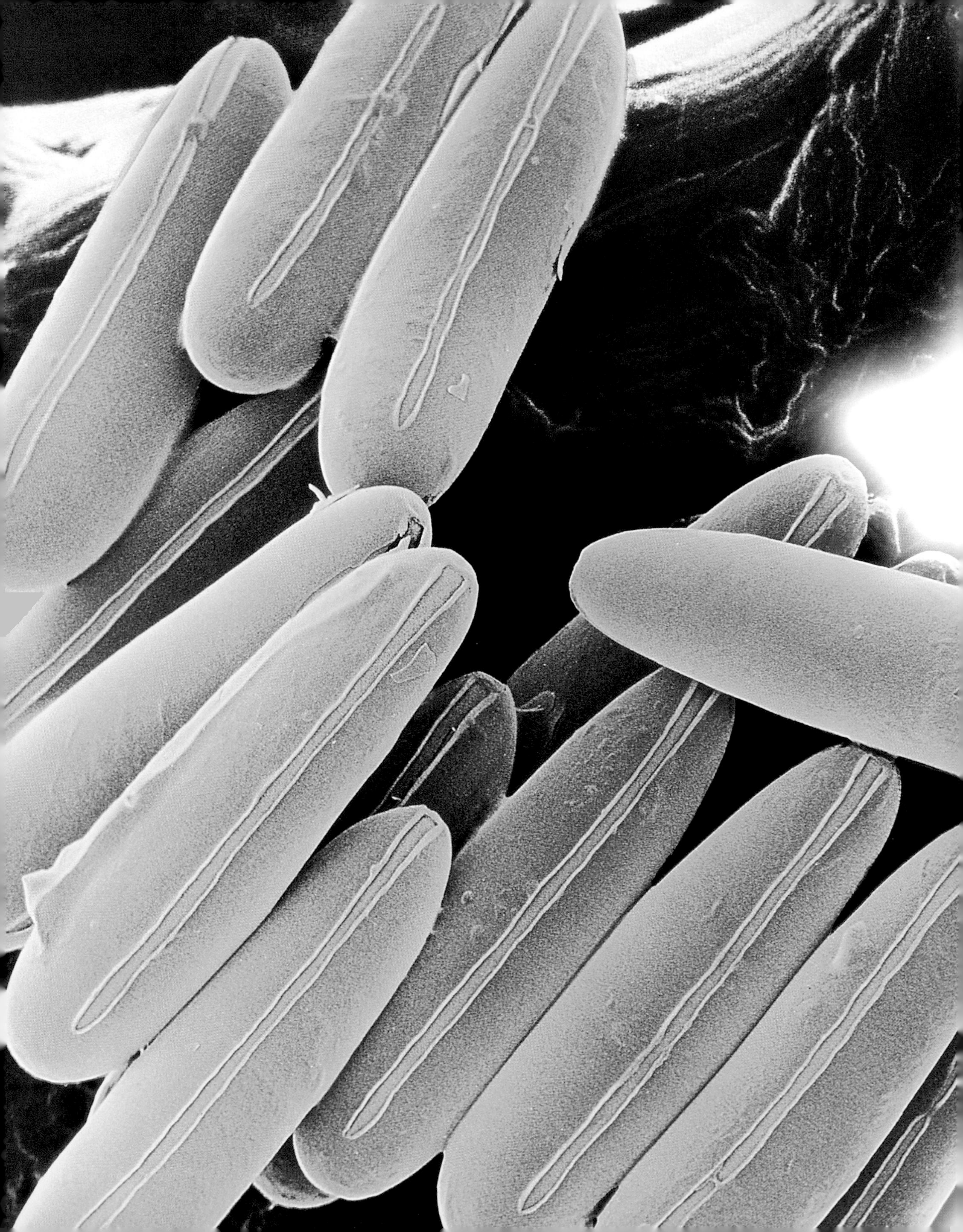

Worm • in the Flesh

Flies will have laid their eggs in a dead body – in the eyes, orifices and any wounds – within hours. The eggs hatch between 8 and 14 hours later, and the stage of development of the maggots can provide a good indication of the time of death.

When a dead body in the open air begins to decompose, the flies arrive, attracted by the smell of putrefaction. They lay their eggs, and after a short time – the specific period depends upon the species – the larvae emerge and begin to feed upon the remains. In due course the larvae pupate and, again after a specific period, appear as new-born flies. If the body has not yet been found, this cycle will repeat itself. It is therefore possible that a knowledge of the behaviour and life cycle of each insect species can provide a fairly accurate estimate of the time that has elapsed since death: accurate to the nearest day or week, however, rather than hours. In more than one case, this evidence has proved important in establishing the dates between which a murder has been committed. It has also sometimes revealed that a body has been subsequently moved from one location to another.

In the open, a body can be invaded by as many as eight successive waves of insects. The first are the bluebottles (Genus *Calliphora*), and the last are the beetles. Bluebottles may have laid their eggs in wounds, on the eyes and lips, and in orifices such as the mouth, nose or vagina within hours of death. This occurs during daylight, and preferably in warm sunlight around the middle of the day; it is much less common – although not unknown – during winter months. Between 8 and 14 hours later, depending upon the air temperature, the eggs hatch, and the first tiny maggots appear. The first stage of development – the first instar – lasts another 8 to 14 hours, and the maggots then shed their skins. The second instar lasts two to three days. By the third instar, the maggot is now a creamy white, and feeds voraciously for some six days. It then migrates some distance from the body and burrows into the ground, where it pupates for some 12 days before emerging as a fly. As bluebottles prefer fresh flesh, the fly is unlikely to return to the dead body.

In the Buck Ruxton case, for example (see "Skull and Bones"), the maggots in the remains were identified as those of bluebottles. The eggs would have been deposited within a couple of days of death, and the life of the largest larvae was unlikely to be more than 12 days. As the remains were found on September 29, and there was no sign of other infestation, they had obviously been lying in the open only since September 17.

Other flies whose maggots may be found on dead bodies are the greenbottle and the sheep maggot fly (Genus *Lucilia*), and the house fly (*Musca*). The life cycle of *Lucilia* is similar to that of the bluebottle, but the house fly, although it feeds upon the dead flesh, lays its eggs in bodies relatively rarely.

Infestation by swarming maggots can raise the temperature of the decomposing body so that it becomes perceptibly warm. This can result in the rapid formation of the fatty substance known as adipocere (see "Gathering the Evidence"). Adipocere forms usually when a body is immersed in water or buried in moist surroundings, and generally takes several months to develop. Nevertheless, cases are known in which maggot infestation has produced adipocere in as little as three weeks, making it appear that the body has lain undiscovered for much longer.

The first instar of a maggot of the greenbottle fly, just two hours after hatching.

A covering of earth prevents some species of fly from reaching a corpse, but the so-called "coffin fly" may burrow into the ground and can even find its way into closed coffins. When the skeleton of "little Miss Nobody" (Karen Price, see "Skull and Bones") was found buried in Cardiff, Wales, in 1989, the police called on the services of Britain's leading forensic entomologist, Dr. Zakaria Erzinclioglu of Cambridge University. Calculating the time it would have taken coffin flies to consume the soft tissues, "Dr. Zak" – as he was known to his colleagues – concluded that

CRIME FILE:

William Brittle

The man's body was in an advanced state of decomposition. The examining pathologist, however, noted the stage of development of the maggots, and was convinced that he had died no more than 12 days before.

William Brittle, who had learnt unarmed combat in the army, murdered Peter Thomas on June 16, 1964. He was sentenced to life imprisonment, largely thanks to the forensic evidence of pathologist Dr. Keith Simpson.

On June 28, 1964, two boys were searching a wood in Berkshire, England, for a decaying rabbit or pigeon, hoping to find maggots for fishing bait. They found a seething mass of fat bluebottle maggots on a mount of loose turf a few yards from their path but, on pulling the turf away, they were appalled to uncover a decomposing human arm.

Pathologist Dr. Keith Simpson was called to the scene, and superintended the disinterment of the corpse. From the degree of decomposition, the police assumed that it had been lying there for six to eight weeks. Simpson disagreed. "At least nine or ten days," he said, "but probably not more than twelve." He calculated, from the maggots' stage of development, that death had occurred on June 16 or 17. "I've seen a body reduced to this state in as little as ten days," he added.

Among the persons reported missing was Peter Thomas, who had disappeared on June 16 from Lydney, on the Welsh border. He was identified from the body measurements, an X-ray of a former broken left arm, fingerprints, and a maker's label in his jacket. He had died from a single violent blow across the throat.

Suspicion fell on a Hampshire resident, William Brittle, who had owed Thomas money – and it was discovered that he had learnt unarmed combat while in the army. He said that he had driven to Lydney on June 16 to repay his debt, and a hitchhiker confirmed that Brittle had given him a lift on his way home to Hampshire that day. At Brittle's trial for murder, the defence brought forward three witnesses who swore that they had seen Thomas in Lydney on June 20 and 21.

Simpson, however, stuck by his findings in court, and was delighted to have them confirmed by an expert brought by the defence. The jury were convinced that the witnesses must have been confused about the dates on which they believed they had seen Thomas, and Brittle was sentenced to life imprisonment.

CASE CLOSED

at least three years had elapsed. Following that, a colony of several generations of woodlice had established itself, and Dr. Zak estimated that this would have taken a further two years – setting the time of burial in 1984 or earlier.

Other insects – beetles, moths and even wasps – can also provide the entomologist with valuable evidence. Such evidence was first used in 1850, when the

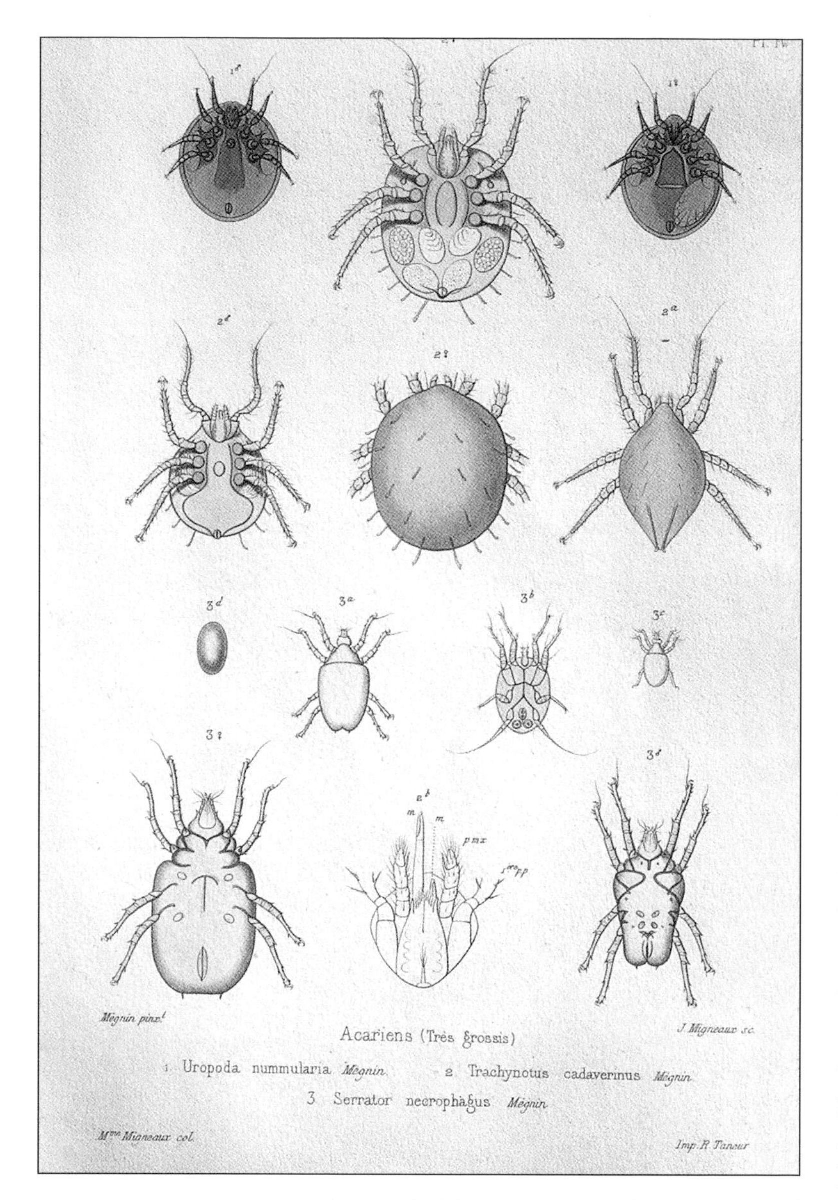

In the later stages of decay, a dead body is likely to be invaded by a wide variety of insects. These drawings of species of mites are taken from La Faune des Cadavres, *a standard work by the 19th-century French entomologist Mégnin.*

mummified body of a new-born baby was discovered hidden beside the chimney of a house. It had become the breeding ground of house moths, and examination of the larvae and adult moths made it clear that the body had been there about two years. This lifted suspicion from recent tenants of the house, and pointed to a woman who had lived there previously, and who had been known to be pregnant. She was found and charged with infanticide. Although it could be shown that she had hidden the body, the charge was dismissed on the grounds that the baby could have died of natural causes.

A similar case turned up in Rhyl, north Wales, in 1960. The mummified body of a woman who had died 20 years previously was found in a locked cupboard. Over the years, moths had consumed the woman's hair, leaving a brief stubble, with clean-cut ends, all over the scalp.

In April 1962, a man's body was discovered in a wardrobe in Denmark. It had been extensively damaged by larder beetles, of which there had been a severe plague during the spring of 1961. The beetles begin to attack a corpse about three to six months after death, when the body fat has turned rancid, indicating that the man had died in the autumn of 1960. Papers found on his body showed that he had been released from prison in August 1960. It transpired that he had gone to live in a friend's apartment, where he died. Instead of reporting his death, the friend had stored the body in the wardrobe. Autopsy revealed that the man had died of natural causes.

A human skull was found in Tennessee in 1985, with a wasps' nest inside it. The nest would have been built, at the latest, in the summer of 1984 and, because wasps only build in dry places, all the tissues, including the brain, must have decayed completely by that time. The subsequent discovery of other parts of the skeleton established that the body had been lying where it was found for at least two years. Death must have occurred in 1983, or even earlier.

Forensic entomology has proved its importance in a number of cases, although circumstances such as weather, temperature, and time of year obviously affect the occurrence of insect infestation, and the expert must take these into consideration. When the case against William Brittle was heard (see above), the expert-witness, Professor McKenny-Hughes, was asked by defence counsel: "Let us suppose that the bluebottle lays its eggs on the dead body at midnight on —". "Oh no," interrupted the entomologist, "no self-respecting bluebottle lays its eggs at midnight. At miday, perhaps, but not at midnight." He added that eight to fourteen hours would elapse before the first maggots began to hatch.

And would these maggots attack the dead tissue at once? "Well," replied the expert, "maggots are curious little devils. Suppose this is a dead body, and suppose you have a hundred maggots here. Ninety-nine will make their way toward the body, but the hundredth little devil – he'll turn the other way." Even the judge found it hard to hide his grin.

FACT FILE

Sometimes the activities of house flies, feeding upon an exposed corpse, can result in deceptive evidence. During a Texas heatwave, the body of a man was found hanging in his home. It had been there three or four days, rapidly decomposing, and the walls and ceiling were spattered with tiny drops of blood. It looked as if the man had been savagely beaten.

An FBI expert examined photographs of the scene, and soon realized that the red-brown spots were not blood, but fly excrement. They had fed upon the body, and then landed on the walls and ceiling with blood on their feet, and excreted the blood they had digested. "One of the things that made that obvious," said the expert, "was a photograph showing a light bulb that had been left on. There wasn't a spot on it. The flies hadn't gone near it because it was too hot for them."

At the French Institut de Recherche Criminelle, specimens of many different species of fly have been assembled for the purpose of identification in the laboratory.

The Finger of Suspicion

Computers have made the classification of fingerprints a much faster operation. Working from an identified central point, the computer can identify the individual features that make each fingerprint unique.

THE REALIZATION THAT AN INDIVIDUAL'S FINGERPRINTS – as well as prints of the whole hand and the sole of the foot – are unique, and so could be used as a means of positive identification, took a long time to develop. Centuries ago, Chinese and Japanese potters impressed their work with a "signature" thumbprint, and practitioners of palmistry predicted an individual's future from the lines of the hand – perhaps they were blinded to the fact that every hand was different by the knowledge that the fate of most people took one of only a few directions.

The surface of the skin on the inner surface of the hands, and on the soles of the feet, differs markedly from that of the rest of the body. From the tips of the fingers to the wrist, the hand is covered with a hornier type of skin that is characterized by a system of papillary ridges. Although these run generally parallel to one another, they occasionally change direction to form clearly defined patterns on various parts of the hand and foot.

The papillary ridges, and the characteristic patterns caused by their change of direction, begin to form during the third and fourth months of the development of the foetus, and after birth no changes occur in their configuration: the only change is in size, the growth of the ridges following the growth of the hands and feet. So far as is known, no two individuals – not even identical twins – have patterns that are exactly alike.

The first steps in the modern use of finger and hand prints for identification purposes were taken during the latter half of the 19th century by two British officials in two far-distant lands.

William Herschel, the grandson of the famous English astronomer, went to India in 1853, at the age of 20. In 1858 he first thought of using a handprint as a signature on a contract, and then began to experiment with fingerprints. He rose through the ranks of the Indian Civil Service and in 1877 was appointed magistrate at Hooghly, near Calcutta, where one of his duties was paying government pensions.

To make sure that pensioners who had died were not being impersonated by others, he began taking prints of the right first and middle fingers on receipts. Soon he had made this the official local policy for all legal documents, and then wrote to the Inspector of Jails and the Registrar-General in Bengal, suggesting that the system should be generally adopted. He stated confidently that the prints did not change with age, and might be valuable in the keeping of criminal records.

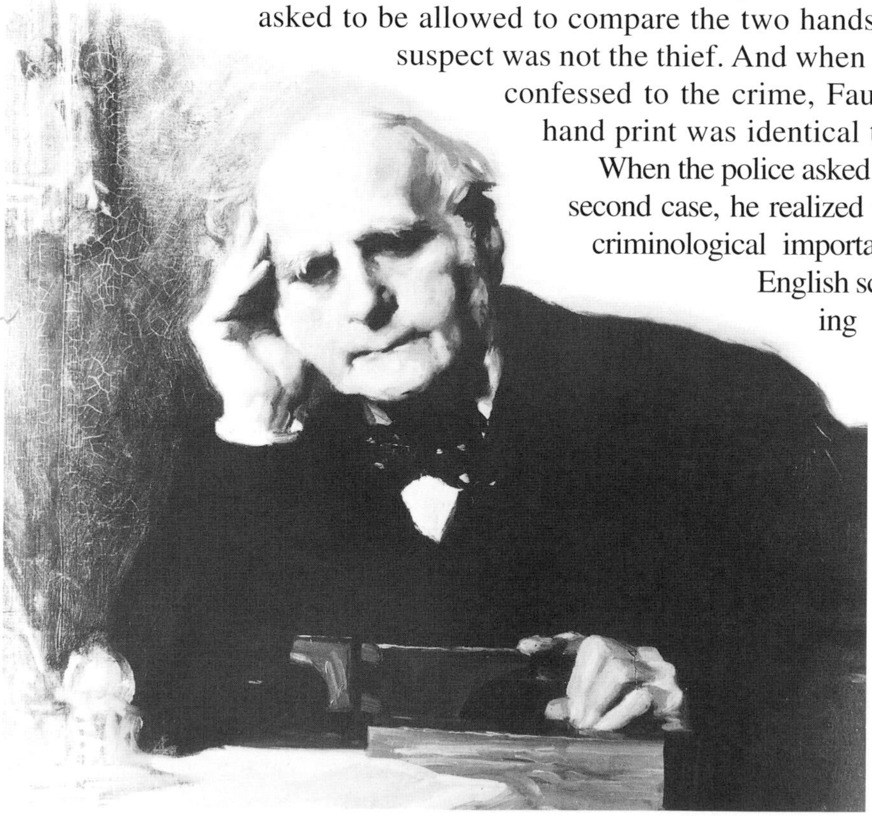

The English anthropologist Sir Francis Galton was the first to develop a system of fingerprint classification, based upon the occurrence of a small triangular area he named a "delta".

A Scottish doctor, Henry Faulds, was working at the same time in Tsukiji Hospital, Tokyo. He noted that documents in the remoter districts of Japan were often "signed" by illiterates with their hand prints in black or red. He wondered whether the patterns of prints varied from race to race, and began to collect specimens. In the summer of 1879, a thief who climbed over the whitewashed garden wall of a house in Tokyo left a sooty hand print behind. When Faulds learnt that the police had already arrested a suspect, he asked to be allowed to compare the two hands, and then announced that the suspect was not the thief. And when another man was arrested, and confessed to the crime, Faulds was able to show that his hand print was identical to that on the wall.

When the police asked Faulds to apply his expertise in a second case, he realized that his observations could have criminological importance, and wrote a letter to the English scientific magazine *Nature*, outlining his theory of what he called "dactylography". He concluded: "There can be no doubt as to the advantage of having, besides their photographs, a nature-copy of the forever unchangeable finger furrows of important criminals."

William Herschel retired from the Indian Civil Service in 1879, and returned to England with the notebooks containing his research. He was mortified to discover that Faulds had pre-empted his discoveries, but he too wrote to *Nature*, and provoked a small amount of controversial correspondence. Further developments in the subject, however, did not take place for some time.

In France, a very different method of criminal identification, making use of body measurements, had been developed by Alphonse Bertillon (see "The Guilty Party"). The English physician and anthropologist, Sir Francis Galton, was unimpressed with "bertillonage", and in 1888 he remembered the Faulds-Herschel correspondence. He contacted Herschel, who sent him his papers, and began to study a way of classifying fingerprints.

Galton discovered that a Polish pathologist named Johann Purkinje had published a paper in the 1830s, in which he described the various patterns in the skin of the fingertips. There were, however, dozens of variables, which would make identifying and comparing prints very time-consuming, and Galton sought for something simpler. Eventually, he hit upon the fact that nearly every print in his collection contained a small triangular area where lines ran together. Galton called this a "delta", and distinguished four basic types: prints with no delta, prints with a delta to the left, prints with a delta to the right, and prints with several deltas. This meant that, if all 10 fingerprints were taken, they could be divided into more than 60,000 classes. In 1892, Galton published his book *Finger Prints*, which detailed his research.

Beginnings of Criminal Investigation

A policeman in Argentina named Juan Vucetich was the first person to put Galton's system into practice. Vucetich had his first success in June 1892, in the case of a mother who murdered her two small children. By 1894, the Argentinian police force had become the first in the world to adopt fingerprints as the principal means of criminal identification. Vucetich described his methods at the Second Scientific Congress of South America in 1901, and within a few years every country in South America had adopted his system.

In Europe and the United States, however, fingerprinting developed in a slightly different direction. Edward Henry was Inspector General of the Bengal Police when he read Galton's book in 1893. With the collaboration of two of his officers, he adopted a system of classification that differed from that of Galton and Vucetich. He described five clearly different types of pattern in the lines of the fingertips: arches (A); "tented" arches (T); radial loops – that is, loops inclined toward the radius bone on the outside of the arm (R); ulnar loops, inclined toward the inner ulnar bone (U); and whorls (W).

After establishing to which of these main classes the print belonged, Henry made a further sub-classification by deltas. He wrote: "The deltas may be formed by either (a) the bifurcation of a single ridge, or (b) the abrupt divergence of two ridges which have hitherto run side by side." He then established the limits of the

The principal types of pattern by which fingerprints are classified:

(1) Plain arch
(2) Tented arch
(3) Loop
(4) Central pocket loop
(5) Double loop
(6) Double loop
(7) Plain whorl
(8) Accidental

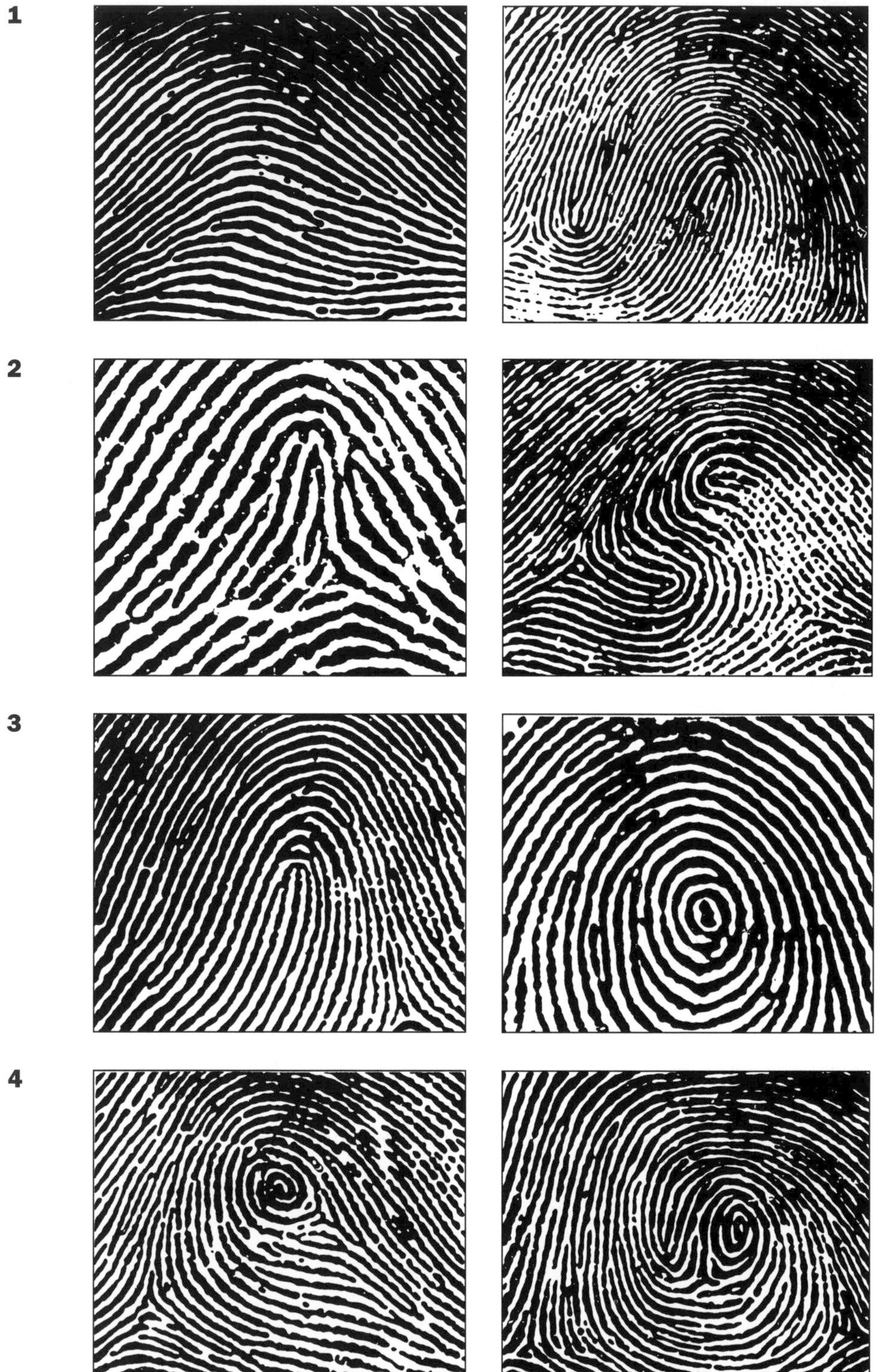

CRIME FILE:

Thomas Jennings

In 1910, fingerprint identification was still a very recent development in crime science. US legal history was made when a burglar who shot and killed a householder was convicted on the evidence of his prints in wet paint.

Clarence Hiller lived with his wife and four children on West 104th Street, Chicago. In the early hours of September 19, 1910, he confronted an intruder on the stairs of the house; two revolver shots rang out, and Hiller died a few moments later.

By chance, police officers about to go off duty had their suspicions aroused by the movements of a man, about a mile away from the Hiller home. They questioned him and, finding a loaded revolver in his pocket, arrested him. He was identified as Thomas Jennings, only recently released from Joliet Penitentiary.

Meanwhile, other officers examined the scene of the murder. The screen on the rear kitchen window had been forced. Also, on the railing of the porch, which was newly painted, they found the clear impressions of four fingers of a left hand. The prints were quickly shown to match those of Thomas Jennings, and he was indicted for murder.

When four expert witnesses testified for the prosecution, the court accepted their evidence, and another step forward was made in the legal history of the United States. Jennings appealed, but the Supreme Court of Illinois ruled on the admissibility of fingerprints on December 21, 1911: "We are disposed to hold … that there is a scientific basis for the system of fingerprint identification, and that the courts are justified in admitting this class of evidence."

CASE CLOSED

delta – from what he named the "inner terminus" to the "outer terminus" – and counted the number of ridges that intersected the line joining these two points.

In 1896, Henry wrote to the Government of Bengal, reporting the ease with which fingerprints could be taken:

> The accessories, a piece of tin and some printer's ink, are inexpensive and procurable everywhere; the impressions are self-signatures, free from possible errors of observation and transcription; any person of ordinary intelligence can learn to take them with a little practice after a few minutes instruction… their characteristics being persistent throughout life, the finger impressions of a child could be used to identify the same person when he had reached middle or old age … and last, the evidential value of identity obtainable from the scrutiny of the prints of two or three fingers is so great that no person, capable of weighing it, would deem it necessary to seek for confirmation elsewhere.

In 1897, the office established in Calcutta by the Bengal government, on the basis of this report, became the first national fingerprint bureau in the world to employ Henry's system. It remains the basis of fingerprint identification to this day.

In 1898, the manager of a tea plantation in northern India was found with his throat cut; his safe and despatch box had been rifled. Among the papers left in the despatch box was a calendar, and on it were two brown smudges, one certainly made by a right-hand finger. Henry had filed the fingerprints of all those convicted of imprisonable offences since he set up his system. The print was soon classified, and identified as that of the right thumb of Kangali Charan, a former servant of the plantation manager. Although he had moved several hundred miles away, he was traced, and a second impression of his right thumb taken. It matched.

Henry was recalled to England in 1901, appointed Assistant Commissioner of the London Metropolitan Police, and given the task of setting up Scotland Yard's Fingerprint Branch. One of his first recruits was Detective Sergeant Charles Collins, who took his appointment very seriously, studying photography for the purpose. He soon had his first success.

On June 27, 1902, a thief broke into a house in Dulwich, south London, and the investigating officer noticed some dirty finger marks on the window sill. Collins photographed a thumb mark and then, with his colleagues, set about comparing the print with those of previously convicted criminals. After a long search, he was rewarded: the print was that of Harry Jackson, who was arrested a few days later.

There remained the problem of persuading a court to accept the evidence. For the prosecution of what was, admittedly, a very minor case, it was decided to retain an experienced counsel, Richard Muir. He spent hours with Collins studying the new system, and in opening the case he explained to the jury how successful the technique had proved in India. Collins then showed how fingerprints were identified, and exhibited his photographs. The jury were intrigued by his demonstration, and found the prisoner guilty – even his defence did not contest the evidence. The acceptability of fingerprints had been established in the English court.

In 1903 Henry was appointed Police Commissioner, and decided to make a sweeping test of fingerprinting in criminal identification. At the Epsom Derby race meeting in May of that year, the police arrested 60 men for a variety of offences; 27 were found to have previous convictions, and their records were available in court the next morning:

> The first prisoner on this occasion gave his name as Green of Gloucester, and assured the interrogating magistrate that he had never been in trouble before … But up jumped the Chief Inspector … and begged their worships to look at the papers and photograph, which proved the innocent to be Brown of Birmingham, with some ten convictions to his credit.

The first case of murder in which fingerprints were accepted as evidence occurred in England in March 1905. Thomas Farrow, the proprietor of a paint shop in Deptford, south London, was found brutally battered to death, and his wife so severely injured that she died three days later. The cash box, which was under her bed, had been rifled, but a right thumb print was found on it. Police inquiries led them to two vicious petty criminals, Alfred and Albert Stratton: the print was found to be Alfred's. Richard Muir led the prosecution of the two brothers and, although the judge remained unconvinced, the jury found both brothers guilty.

J. Edgar Hoover was appointed Director of the Federal Bureau of Investigation in 1924.

Fingerprinting in the United States

In 1904, the World's Fair and Exposition was held in St. Louis, Missouri, and John Ferrier of the London Fingerprint Branch was among the police officers sent to guard the British Royal Pavilion. While there, he gave a number of lectures on the Henry system, and a mere ix years later, the acceptance of fingerprints in a United States murder case was established with the conviction of Thomas Jennings in Chicago in 1911.

In the early 1900s, the United States Department of Justice took the decision to allocate a sum "not to exceed $60" in order to establish a system of fingerprinting at Leavenworth prison in Texas. In 1905, Sing Sing and other prisons in New York State began to employ fingerprinting, and in the following year the St. Louis police also adopted the technique. At about the same time, the Army, Navy and Marine Corps began to fingerprint both their enlisted men and their officers. Before long, it became clear that some means of coordinating all these separate records was needed, and the Department of Justice undertook the task. Unfortunately, they transferred the job of cataloguing to Leavenworth, where it soon transpired – though it should have been no great surprise – that the convicts employed were on occasion altering the records to their own advantage.

Since 1896, the International Association of Chiefs of Police (IACP), which comprised the heads of police departments in most major cities of the United States and Canada, had maintained a National Bureau of Criminal Identification. This bureau was based at first in Chicago, and later in Washington DC. The IACP campaigned vigorously for fingerprint records to be centralized. What was to become the FBI was established within the Department of Justice. It was not until 1924, however, when J. Edgar Hoover was appointed Director of the Bureau, that the cataloguing of some 800,000 records, which had accumulated higgledy-piggledy in storage, was finally undertaken.

Hoover quickly realized the importance of keeping records of persons who had not been involved in any crime. They would be invaluable in tracing missing persons, identifying remains mutilated in disasters, reuniting victims of amnesia with their families, and clearing innocent people of suspicion. Hoover's developing paranoia in later years no doubt led to misuse of these records, but nobody could dispute their practical value. There are currently more than 200 million prints, representing over 68 million people, on file at the FBI.

A Magna-Brush being used to develop latent fingerprints with magnetic powder. The surface is that of a polystyrene tile, on which it would be impossible to develop prints by other dusting techniques.

Latent Prints

The earliest successes with print identification concerned those that were visible: in blood, or a similar medium, or impressed into a plastic surface. Researchers soon discovered, however, that invisible – "latent" – prints could be detected on almost any smooth surface.

Latent prints are formed by minute traces of sweat, either from the fingertips themselves, or after unconscious contact with the face or another part of the body. They weigh anything from 4 to 250 micrograms, and are about 99 percent water, the remaining 1 percent being a complicated mixture of substances that varies, not only from individual to individual, but from hour to hour in the same individual. The persistence of such a print can depend upon a variety of factors, but it can be almost permanent: latent prints have been taken from objects found in ancient tombs.

Latent prints can be "developed" in a number of ways. The basic technique is to dust with a very fine powder – using either a camel-hair brush, or an "insufflator": a device very like a scent spray. In the early days, a mixture of mercury and chalk ground finely together was used, but this has been superseded by other materials.

Prints on glass, silver or dark surfaces are dusted with a light grey powder; while prints on light-coloured non-absorbent surfaces are dusted with black powder. A device known as a Magna-Brush uses magnetic powders: some small particles adhere to the print, and any excess can be removed from the surrounding areas with the magnet. Obviously, this can be used only on non-ferrous surfaces. Coloured and fluorescent powders are also employed.

For many years, these developed prints had then to be photographed; but more recently the procedure has been to "lift"

Fingerprints developed on the inside of a neoprene glove by the Magna-Brush.

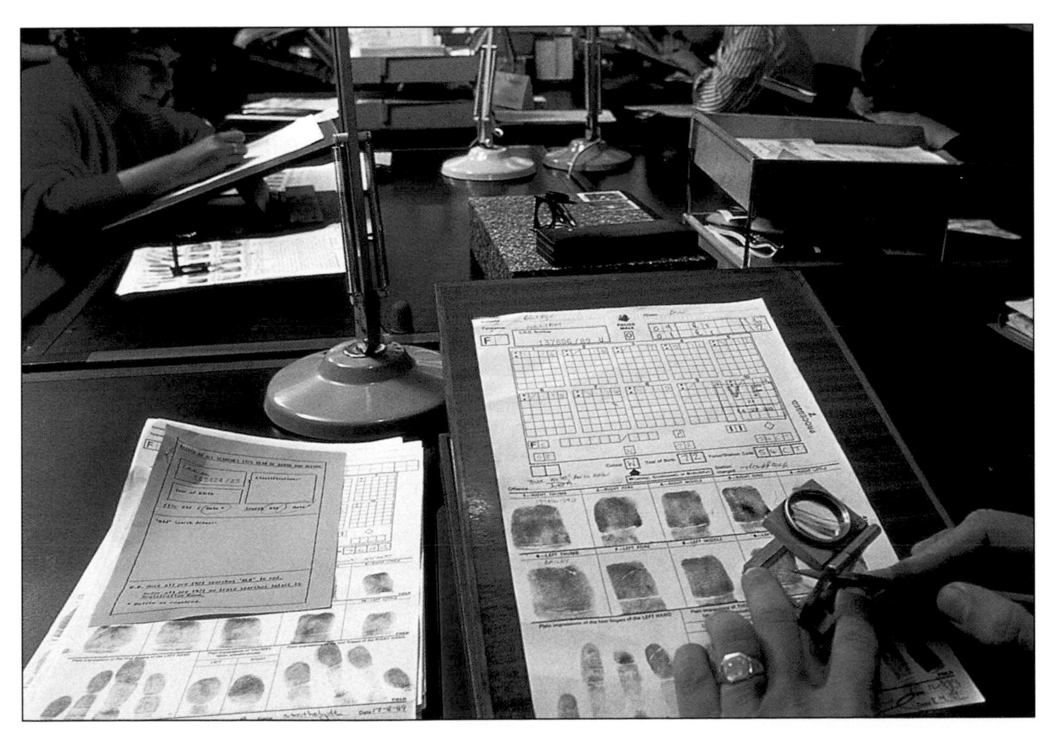

For many years, the identification of a fingerprint found at the scene of a crime took hours of laborious searching through the records. Here an officer at London's New Scotland Yard examines a record for comparison.

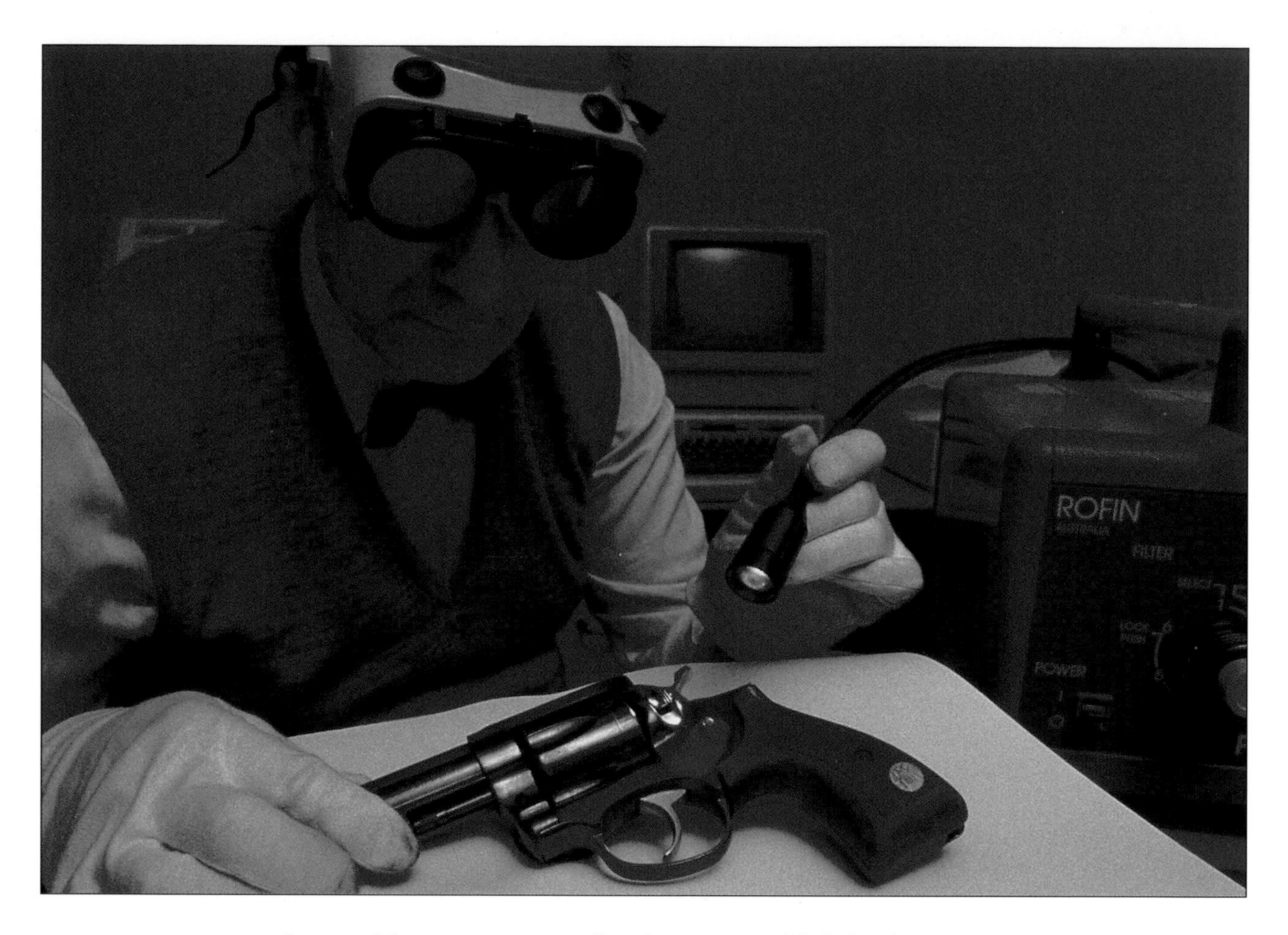

A forensic scientist, wearing red-coloured goggles to cut out extraneous light, searches with a laser for fingerprints on a revolver found at the scene of a crime.

them with transparent adhesive tape, which is then mounted on a transparent backing, or a card of a suitable colour.

Prints on porous surfaces such as cardboard or wood must be revealed in a different way. For many years the standard methods were silver nitrate, which reacts with the salt in sweat; or iodine vapour, which reacts with grease. Then, in 1954, it was discovered that the amino acids in sweat were sufficient to react with a substance called ninhydrin. Sprayed with a dilute solution of ninhydrin in acetone, and then dried in an oven at 176°F (80°C), a piece of paper or card can reveal purplish latent prints. This is now a standard method of print development. Similar techniques employ other dyes that react with proteins.

Prints on human skin – particularly valuable evidence in cases of rape – can be lifted with the very high-gloss paper known as Kromecote, or illuminated by specialized X-ray techniques. However, they seldom survive for more than two hours.

Searching for latent prints can be a time-consuming process, as the entire scene of the crime must be examined. A remarkable, and very useful, discovery

CRIME FILE:

Peter Griffiths

Before fingerprints could be catalogued on computers, the task of identification was a daunting one. More than 46,000 prints had to be collected and examined before the brutal murderer of a little girl was identified.

At midnight on May 14, 1948, three-year-old June Devaney was asleep in her cot in the children's ward of Queen's Park Hospital in Blackburn, northern England. At 1.20 AM the night nurse discovered that the little girl was no longer in her cot; on the floor beside it stood a large glass Winchester bottle*. The nurse raised the alarm, but a speedy search through the hospital and its surroundings revealed nothing, and at 1.55 AM the police were called. At 3.17 AM June Devaney's body was found near the boundary wall of the grounds: she had been brutally raped, her left buttock was deeply bitten, and she had been battered to death against the stone wall.

Detective Inspector Colin Campbell, of Lancashire Constabulary's fingerprint department, lifted 10 prints from the Winchester bottle that were not those of any of the hospital staff. At a police conference on May 18, he said that these consisted of a left thumb print, the four fingers and palm of the left hand, two fingers of the right hand, and three other partial prints. Because of the large span of the prints, the clarity of ridge detail and the absence of coarseness or signs of scars, he thought it likely that they had been made by a well-built young man, but one who had done little heavy manual work.

A search of all the fingerprints filed throughout the country would take a long time, at the end of which it was still possible that the prints would remain unidentified. Realizing this, the police decided to take the prints of every male over the age of 16 who was known to have been in Blackburn on May 14 or 15 – nearly 50,000 of them. The mayor of Blackburn set an example by being the first to volunteer his prints, and police officers set to work, taking 500 sets of prints a day.

On July 18, after two months of overtime work, the operation was temporarily suspended while Inspector William Barton and his staff checked through more than 40,000 records, none of which had yet produced a match. They resumed their house-to-house inquiries on August 9, and on August 11 they took the prints of 22-year-old Peter Griffiths, a former soldier in the Guards. On the afternoon of the following day one of Campbell's team, wearily going through the latest batch, stopped, looked again, and cried out "I've got him! It's here!" The number of Griffiths's set of fingerprints was 46,253.

Griffiths confessed, and was hanged on November 19 of the same year.

CASE CLOSED

** A Winchester bottle is a cylindrical bottle for transporting liquids. The term originally applied to containers holding a bushel, gallon, or quart, according to an obsolete system of measurement with standards kept at Winchester, Hampshire, England.*

has been that prints exposed to the fumes of "superglue" (cyanoacrylate) will show up white on darker surfaces. This is particularly valuable for examining enclosed spaces, such as cupboards or car interiors. The revealed prints can be dusted, and photographed or lifted.

Researchers in a Canadian laboratory made another important accidental discovery, when they found that laser beams would show up latent prints. Unlike powders or chemicals, lasers do not affect the object on which the

prints are found, and rather oddly they seem to be more effective on older evidence. In 1975, the United States Department of Justice sought a deportation order against Valerian Trifa, a former archbishop of the Romanian Orthodox Church, on the grounds that he had concealed his former membership of Romania's pro-Nazi party, the Iron Guard. Trifa denied the charge, but in 1982 the West German government found a postcard that Trifa had written to a high-ranking Nazi official. The Germans refused to allow the FBI to employ any destructive process, but a laser quickly revealed Trifa's thumb print on the card, and he was deported in 1984.

Modern digital technology has made it possible to enhance prints that were otherwise too diffuse to analyze. At one time, the cost of a suitable computer was beyond the means of most police forces but, now that personal computers are within the reach of everyone, forensic science laboratories have adopted the technique.

Valerian Trifa, former Romanian archbishop, who concealed his membership, nearly 40 years earlier, of the pro-Nazi Iron Guard. In 1984 the US Department of Justice was able to secure his deportation on the evidence of his fingerprint on a postcard to a high Nazi official.

Further Developments

On a standard fingerprint form, two sets of prints are taken. The first – "rolled prints" – are taken in 10 numbered compartments. Each finger is inked, then rolled completely from edge to edge, so that patterns that extend around the curve of the finger are recorded. Then the same 10 prints are taken as "plain" impressions, without any rolling action. This is principally to ensure that the same prints are taken in the correct sequence – there have been occasions when criminals have apparently been "helpful", holding out their fingers in the wrong order, or even contriving to have the same hand printed twice.

Where prints are found at a scene of crime, the prints of all persons who might have had access to the area must be taken, including those of all examining officers, to eliminate these from the inquiry.

However, it is not practical to search thousands of files for a matching set of ten prints. When only a single print is found at the scene of crime, it is almost impossible to find a matching print on the standard file cards. A way of classifying each single print was needed. At Scotland Yard, beginning in 1927, Detective Chief Inspector Harry Battley introduced a more comprehensive filing system. He designed a special fixed-focus magnifying glass with a glass base engraved with seven concentric circles with radii from 0.12 inches (3 millimetres) to 0.6 inches (15 millimetres), identified by the letters A to G. By centring the glass over a fixed point in the fingerprint, usually the core (what appears to be the centre of the pattern), it was easy to classify the delta by the circle in which it appeared.

Old fingerprint records on card can now be scanned and transferred to computer by means of a simple – and relatively cheap – hand-held digitizer. The computer data-base can analyze and compare millions of fingerprints, and obtain a match within minutes.

A separate collection of prints was assembled for each digit. Each of these ten collections was then sub-divided into nine classifications of print type: arch, tented arch, radial loop, ulnar loop, whorl, double loop, pocket loop (with a tiny whorl at its centre), composite and accidental (similar to a double loop, but with one loop enclosing a tiny pocket). Within these sub-divisions each print was then filed according to the classification of its delta. This Single Print System is now the basis of all fingerprint analysis.

English law requires a minimum of 16 matching characteristics to establish identity in the ridge pattern of finger or palm prints, or 10 + 10 in two prints: the probability of two persons having the same fingerprint is 1 in 10,000,000,000,000. In France, 17 matching points are required. Only 12 are required in Greece, Switzerland or Spain; and in Sweden 10 are acceptable. The United States abandoned any formal standard requirement in 1973.

In the 20th century, the number of stored fingerprint records became so great that the task of physically checking through them, no matter how detailed the classification system, grew daunting. Many police forces gave up dusting for prints in relatively minor cases such as thefts from homes and cars. Even in serious cases the search for prints was considered scarcely useful, and was carried out more as a public relations exercise, in order

CRIME FILE:

Richard Ramirez

A California state fingerprint expert described it as "a near miracle", when newly computerized prints identified the multiple rapist and murderer who had been dubbed the "Night Stalker".

Richard Ramirez, the "Night Stalker" of the Los Angeles suburbs, whose fingerprints were raised by computer in a "near miracle".

Between June 1984 and August 1985, the inhabitants of the suburbs of Los Angeles, California, went in fear of a serial killer they dubbed the Night Stalker. Often, he would break into a house in the early hours of the morning, summarily shoot any adult male with a bullet to the head, then rape his female partner. In March 1985, for example, he shot pizza-parlour owner Vincent Zazzara, repeatedly slashed his wife Maxine, and gouged out her eyes. Occasionally, he abducted his victim and, surprisingly, on or one or two occasions he let her live. These survivors gave police partial descriptions of their assailant – he was thin, with dark curly hair, staring eyes, and a mouthful of rotting teeth, and he stank – but there was little else to go on.

On the night of August 5, 1985, the Night Stalker attacked Christian and Virginia Peterson and, although badly injured, both survived and were able to give a good description of him. An artist's sketch was featured in newspapers and on television the following day.

On August 17, he broke into the Lake Merced home of William Carns, shot him three times through the head (Carns survived, but with permanent brain damage), and raped his fiancée twice. He told her: "You know who I am, don't you? I'm the one they're writing about." He spared her life and, despite her terrible experience, she saw him drive away in a battered orange Toyota.

The car had been spotted earlier by a keen-eyed teenager, who notified the County Sheriff's department. The car, which had been stolen, was found abandoned in a parking lot. It was taken away for examination, and laser scanning revealed an acceptable fingerprint.

Only a few days before, the state criminal computer at Sacramento had been updated with all the prints on file of persons who had been born since January 1, 1960. Within minutes, the print was identified as that of Richard Ramirez, a 25-year-old drifter from El Paso, twice convicted, several years previously, for car theft. He had been born on February 28, 1960.

Photographs and a description of the wanted man were immediately circulated. He, meanwhile, was on a trip to El Paso, and returned to Los Angeles unaware that his likeness was on the front page of every newspaper. When he entered a liquor store he was quickly recognized, and fled, pursued by local people. Eventually he ran up to a police patrol car, gasping: "Save me, before they kill me!"

On September 20, 1989, Ramirez was finally convicted of 13 murders and 30 other felonies, and sentenced to death.

CASE CLOSED

The British National Fingerprint Gallery housed close on 5 million records. These have now been digitized, and the laborious manual search shown here is no longer necessary.

to reassure the public that the police were doing a thorough job. Moreover, if the prints were of someone who was not already on file, the search would prove futile. In some major crimes, it has proved more fruitful to take the alternative course, and carry out mass fingerprinting among the entire population of a district, when it was believed that the perpetrator might be living among them.

The search for filed prints, and consequently the value of taking prints from the scene of the crime, has now been made immeasurably easier by the introduction of computer databases. Once the classification details of a print have been measured, a database containing millions of records can produce a match in minutes. There still remains the awesome task of entering the records one by one, but this can be speeded up by the use of scanners, which can also enter a digital copy of the print itself.

The process of taking new prints can also be accelerated with a scanner, which scans each finger electronically and produces a standard print card. Security systems that use a similar principle, recognizing the thumb or forefinger, are already installed in many government, commercial, and domestic premises. The next step is to link the scanner directly to the database.

Problems still arise because different police forces have often purchased different models of computer, or different types of software, which make it impossible for one database to communicate with another. Steps are now being taken

Many criminals hope to avoid the leaving of any fingerprints by wearing surgeon's latex gloves. However, they often discard them close to the scene of the crime and, as the photograph on page 119 shows, clear prints can be obtained from the inside of a glove.

to build up national – and international – databases. The FBI and Interpol are already a long way down this road, and the British police expect to have a national database completed in 2001.

The FBI's National Crime Information Center (NCIC) in Washington DC deals with more than 30,000 fingerprint inquiries every day, and can electronically transmit the discovered print to the appropriate computer terminal. Police cars are increasingly being fitted with mobile computers, enabling comparison of prints to be made on the spot.

Some Remarkable Successes

Fingerprints can be discovered in the most unexpected circumstances. In one English case, a woman was attacked by an unknown assailant; she successfully beat him off, and was taken to hospital for treatment. In examining her face and lips, the doctor noticed a fragment of skin wedged between two of her lower teeth. The woman recalled biting at the man's fingers, and an immediate close examination of the fragment by the police revealed a whorled pattern.

Some hours later, a man with an injured finger was taken in for questioning by the police. He claimed that he had injured it at work, but his fingerprints were taken, and the print of his left middle finger showed that a piece of skin had been torn from the centre of the tip. The fingerprint expert was able to prove that the fragment taken from the woman's teeth exactly fitted the missing area.

In another case, a man wore a pair of surgical gloves to break into a post office in Manchester, England, believing that they would leave no identifiable traces. However, he carelessly stripped them off as he left. Turning the gloves inside out, the police obtained a complete set of prints from their inner surface.

In an American case, the body of a woman who had been tortured, raped, and then murdered, was found floating in a stream. Her clothing was sent to the FBI laboratory for examination, and one expert noticed an unusual pattern in the fabric of her tights. When he examined it closely, he discovered that a fin-

CRIME FILE:
Bertie Manton

Who was the murdered woman? Clues led to her possible home, but there was no sign of her fingerprints – until a film of dust on the last bottle, in the furthest corner of the cellar, yielded its secret.

On November 19, 1943, two sewer workers found a half-submerged bundle in the river Lea in Luton, in southern England. Inside, they discovered the body of a woman. She was naked, and had been killed by a blow to the head with a blunt instrument. The autopsy revealed that she was aged 30 to 35, and pregnant. There were no teeth in either jaw, so she presumably wore dentures – but there was no sign of these, nor of anything else that would identify her. Her fingerprints were taken, but matched nothing in the police files.

The police carried out extensive house-to-house inquiries, but the injuries to the woman's head made the post-mortem photographs of her difficult to recognize. At 14 Regent Street, the two sons of Bertie Manton assured police that the photographs were not of their mother, who in any case, they claimed, was alive.

Months passed without a lead. Discarded pieces of women's clothing were collected, mostly from rubbish dumps. On February 21, 1944, a piece of padding from a woman's black coat was discoverd, with a dyer's tag bearing the number V12247. The Sketchley Dye Works in Luton confirmed that they had taken an order on March 15, 1943, to dye a woman's coat black. The woman's name was Irene Manton, of 14 Regent Street.

It emerged that Irene Manton had not been seen since November 18, 1943. Her husband Bertie was questioned: he said that the photograph was not of his wife, who had left home after a quarrel and was now living in London.

Superintendent Frederick Cherrill, Scotland Yard's leading fingerprint expert, searched the Manton home to see if he could find any fingerprints to match the dead woman's. In a cellar, he found a shelf crammed with empty glass jars and bottles. The walls of the cellar were grimy with dust, but there was no dust on the bottles. They had obviously been scrupulously cleaned.

Bottle after bottle was carefully examined, and discarded. Then, "lurking in the shadows of the remotest corner…I came across the last bottle.... On its sloping shoulders was a film of dust. I tested it ...I found a thumbprint which corresponded with the left thumb impression of the dead woman."

Bertie Manton confessed, and was sentenced to life imprisonment.

CASE CLOSED

gerprint had somehow been burnt into the tights. Reading the details of the murder, he found that the woman had been tortured with a hot knife. Evidently the murderer had wrapped the tights round the knife to protect his hand. The fabric had melted, and he had impressed his fingerprint in the molten material. This evidence eventually led to a conviction.

Some criminals, realizing how easily their fingerprints may identify them, have made painful efforts to destroy them. In 1941, Roscoe James Pitts

CRIME FILE:

James Smith

The only evidence of murder was a single hand and arm, already decomposing. Forensic experts meticulously reassembled flakes of skin to obtain fingerprints that successfully identified the missing victim.

On the afternoon of April 25, 1935, a holiday crowd was attracted to an aquarium in the suburbs of Sydney, Australia, by the spectacle of a tiger shark that had been captured alive by fishermen eight days before. Suddenly, the creature began to thrash wildly in its tank, and thenspewed up a tattooed human arm. The Scottish pathologist Sir Sydney Smith happened to be in the city at the time, and he was asked to examine the arm. "I found that the limb had been severed at the shoulder joint by a clean-cut incision, and that after the head of the bone had been got out of its socket the rest of the soft tissues had been hacked away." The condition of the arm's blood and tissues further suggested that the amputation had taken place some hours after death, and the arm had then probably been dumped at sea. The shark, it turned out, had merely taken advantage of a convenient snack.

Forensic experts carefully removed the skin of the fingers in flakes, and reassembled them so that prints could be taken, an operation that took several weeks.

The prints were faint, but sufficient to establish the identity of the victim as James Smith, a suspected drug runner.

The remainder of Smith's body was never found, and the police were unable to secure the conviction of any of the suspects in the murder, when a justice of the Australian Supreme Court ruled that "a limb does not constitute a body".

CASE CLOSED

(aka Robert Philipps) had the skin surgically removed from his fingertips, which were then sewn on to the skin of his chest until they healed. There was then no ridge pattern on the tips. However, both his original print card, and the prints taken when he was again arrested, included portions of the skin patterns below the first joint, which successfully identified him.

In 1990, police in Miami, Florida, arrested a suspect in a drug case, whose fingertips were heavily scarred. He had sliced the skin from his fingertips, cut it into small pieces, and transplanted the pieces on to other fingers, making himself a human jigsaw puzzle. FBI expert Tommy Moorefield spent weeks solving the problem. Working with enlarged photographs, he cut them into pieces, and gradually matched up the patterns of lines to produce prints that corresponded with those of a man wanted in another drug case.

Detective Chief Inspector Tony Fletcher, head of the police Fingerprint Bureau in Manchester, England, came across a most unpleasant case. A thief had broken into a house and stolen many valuable items, leaving a trail of havoc and glove prints behind him. However, as often happens, he had also left a large bowel movement on the kitchen floor, and adhering to it was a makeshift piece of toilet paper. Fletcher wrote: "I carefully removed the paper and took it back to the Fingerprint Bureau, where I subjected it to chemical tests…There developed on the paper was a perfect right middle-finger print."

Dead bodies can be identified by their prints, but sometimes circumstances make this difficult. Badly shrivelled or wrinkled skin must be restored to its former shape by the injection of a chemical known as tissue restorer. When bodies have been a long time in water, the outer layer of skin, the epidermis, may be macerated so that the pattern of lines is no longer visible, or even partially destroyed by friction with sand or gravel. The epidermis may also be destroyed by fire. In these cases, the epidermis can be delicately removed, and prints obtained from the second layer, the dermis.

If the patterns are intact, but the skin is too soft to be printed, it can be removed in a similar way, and slipped over the examiner's own finger like the finger of a glove. Alternatively, it can be turned inside out, and a reverse print made of the ridge detail.

In many recent cases of murder, the perpetrator has taken great pains to wipe clean every item or surface that might have been touched, but it is almost impossible to remember every little detail, and meticulous examination frequently reveals a single print that has been left untouched (see page 127).

Palm Prints

After Superintendent Battley had successfully introduced the Single Fingerprint System at London's Scotland Yard in 1927, he and his assistant, Detective Inspector Frederick Cherrill, turned their attention to palm prints. They realized that prints of the rest of the hand were as distinctive as fingerprints, but in the beginning palm prints were filed only with their related fingerprints, and could not be separately identified.

Fortunately, Cherrill was soon able to prove the value of palm prints. In 1931, a burglar named John Egan was active in the northwest suburbs of London, and in one break-in he left his hand print on a glass tabletop, as well as sufficient fragmentary fingerprints to identify him. When he was arrested, Cherrill took his palm print, and convinced him that this evidence was so strong that Egan pleaded guilty.

Because he was not called to give evidence in this case, Cherrill was unable to establish the acceptability of palm prints in law for several years. Then, in 1942, intruders attacked an elderly pawnbroker named Leonard Moules at his premises in Shoreditch, London. They beat him to death with a revolver butt, and rifled his safe. Cherrill found a single palm print on the inside of the safe's door, but had no way of tracing it in his records. However, police inquiries led to the arrest of George Silverosa, whose palm print was found to match. He admitted his involvement in the crime, but blamed his accomplice Sam Dashwood for the murder. At their trial, both men declined to give evidence, and were found guilty and sentenced to hang. This was the

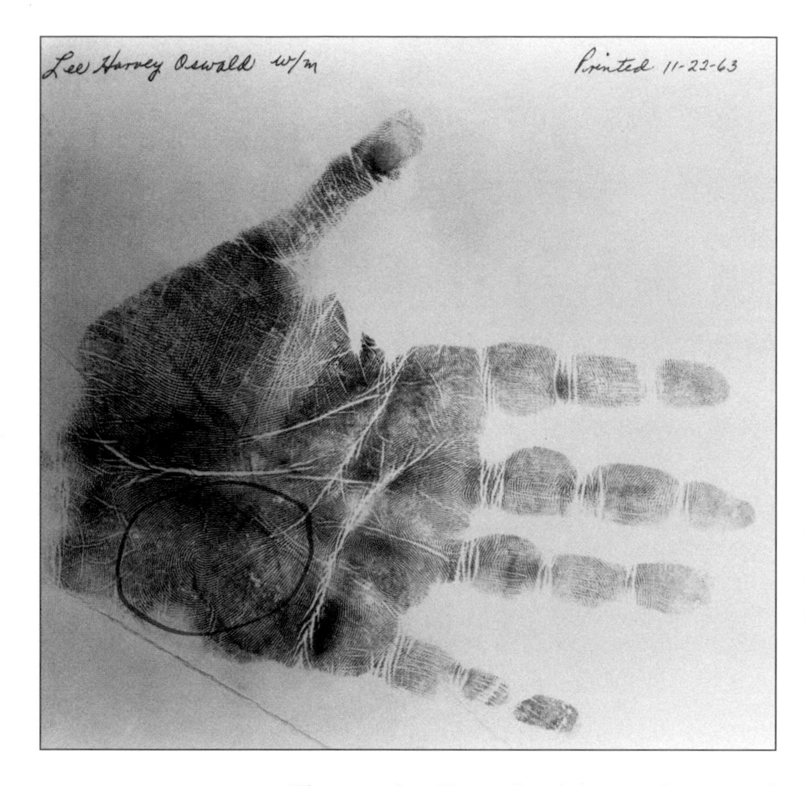

Records, not only of fingerprints but of palmprints, are now being kept and classified. They can prove as distinctive as fingerprints and, like fingerprints, can reveal distinctive identification patterns. The print above is that of Lee Harvey Oswald, the assassin of President Kennedy in 1963.

first case of a palm print being admitted as evidence in English law. Others soon followed.

Glove Prints

Criminals have frequently resorted to wearing gloves, to be sure of leaving no fingerprints. The confidence that this engenders, however, is often misplaced. We have already seen how a set of prints was obtained from the inside of a pair of surgical gloves; on another occasion, a burglar took a new pair of rubber gloves to the scene of crime, tore off the paper band surrounding them – and left a perfect set of his prints on the paper.

These were lucky breaks for the investigator. Gerald Lambourne, who spent many years in Scotland Yard's Fingerprint Branch, rising to become its head, devoted the latter part of his service to a study of glove prints. In 1984 he wrote:

> From the constant handling of objects such as door-knobs or handles, banisters, shop doors, rails on buses or trains, on which people have deposited large quantities of perspiration, a glove soon becomes impregnated with grease and dirt. Even during the putting on and removal of a glove, a layer of perspiration is placed on it… As with fingerprints, the grease-deposit from the glove can be made visible by the skilful application of a fingerprint search powder. The layer of grease left by the glove is by its mere nature not as strong and dense as a fingerprint, but it is nevertheless detectable, recordable, and in certain cases can be as reliable as a fingerprint when it comes to identification.

Lambourne recognized that, when a particular type of glove was manufactured in its hundreds of thousands, it could be difficult to prove its individuality. This objection, he realized, could be brought particularly against ordinary household rubber gloves, which have a standard grip pattern on their tips. He examined hundreds of these gloves, and discovered that, in their manufacture:

> Slight uneven adhesion of the latex [to the mould] can mutilate what should be a regular pattern. Air bubbles can occur in the pattern area…or a piece of latex

.W.D.C.
MATE COUNT

DATE ______________________________

LOCATION ______________________

AT THE TIME OF ______________

THERE WERE __________ INMATES

____

CORRECTIONAL OFFICER

CRIME FILE:

Michael Queripel

In a landmark case in London, police took nearly 9000 palm prints in a mass operation, before identifying a brutal murderer.

On the night of April 29, 1955, Alfred Currell reported to the police that his wife Elizabeth had not returned from walking her dog on a golf course at Potter's Bar, north of London in England. The dog had returned home alone.

At dawn the following day Mrs. Currell's battered body was found. She had been killed with a heavy iron tee-marker from the 17th tee close by. There were signs that the attacker had attempted rape, but the pathologist, Dr. Francis Camps, soon verified that sexual assault had not taken place. On the tee-marker, the police found a partial palm print in blood.

At that time, Scotland Yard had some 6000 palm prints on file, but no match could be found. It was decided that the only course would be a mass hand printing operation.

Toward the end of June, house-to-house teams began to take the prints of all males who lived or worked in the area. The teams checking the prints were ordered to work on the task for only a week at a time, with a return to normal duties for a week in the interim, to avoid eye fatigue. A backlog soon built up, and by the middle of August nearly 9000 prints had been taken. Then, on August 19, print no. 4605 was identified. It had been taken some weeks earlier from 18-year-old Michael Queripel.

Queripel at first maintained that he had come across Mrs. Currell's body while taking a walk on the golf course, but later confessed. He was found guilty, but avoided the death sentence by a few days – at the time of the murder he had been only 17 years old.

CASE CLOSED

from a previous glove can contaminate a portion of the pattern. All of these factors can be detected in a glove print.

By 1971, Lambourne was satisfied that a glove print could be positively identified. On January 29, an alarm bell at premises in London's Pimlico brought police to the scene, and they arrested a man found climbing over a wall at the back. He protested his innocence, but it was soon discovered that a window on the premises had been broken, and a fingerprint officer discovered a glove print on pieces of the glass.

Lambourne examined the print, and satisfied himself that it had been made by a left-hand glove with a suede finish, the surface of which had been damaged. The gloves carried by the prisoner were produced; they were made of sheepskin with a suede finish, and the surface damage on the left-hand glove exactly matched the print. Although the man pleaded guilty, Lambourne was allowed to explain his evidence to the court; it was accepted, and became another precedent of English law. In later cases, Lambourne successfully presented evidence about gloves made from leather, PVC, rubber and cotton twill, and his findings were adopted by police investigators throughout the world.

FOOTPRINTS

Prints of naked feet are not often found at the scene of the crime, but there have been cases in which a burglar has taken off his shoes and socks, and placed the socks on his hands, so as not to leave any fingerprints – not realizing that latent prints from feet are just as identifiable. Some police forces also keep records of ear prints, such as those left by burglars who have listened closely at a window before breaking into a house.

A latent print left by a naked foot can provide positive identification of an individual, but prints left by shoes and boots can be almost equally important. Footprints found at the scene of crime may be indented in mud, sand, or other materials, in which case a cast in plaster of Paris or silicone rubber is made of them, or, if they are visible on a dusty floor, they can be photographed.

Boot and shoe prints at the scene of a crime can prove to be a valuable piece of evidence. Here, the occurrence of a sharp frost later in the night has helped to highlight the print.

In 1945, a succession of break-ins began in New York State. In the first two, the thief left fingerprints, but no matching prints were found in the police files. A year later, police found a print of an unusual overshoe in the garden of a house that had been burgled, and they took a cast of it. In August 1947, the print of a sneaker was discovered, and a few days later another identical print was found, together with fingerprints that matched those taken in 1945.

The breakthrough came in November 1947. Following a robbery at a gas station in Seneca Falls, the police were given the licence-plate number of a car seen parked nearby. They traced the owner, who admitted his part in the robbery, but implicated his uncle, who lived nearby. At the uncle's house, the police

CRIME FILE:

A Postman in Cairo

Experienced native trackers can tell much from a few prints in the sand. Bedouin skill led to the detection of the murderer of a local postman in 1920s Egypt.

In the 1920s, while the Scottish pathologist Sir Sydney Smith was serving as medico-legal adviser to the Egyptian government, the body of a local postman, shot through the head, was found in the desert on the outskirts of Cairo. Although the surrounding sand bore no identifiable marks, the commandant of the city police decided to enlist the aid of Bedouin trackers. "They could without difficulty spot the tracks of different persons they knew," wrote Smith, "and could tell whether a person...was running or walking, whether loaded or free, and so on."

The Bedouin trackers found prints of a man in sandals that led from a spot some 40 yards (36.4 metres) away, where, they said, he had knelt. Close by they found an empty .303 rifle cartridge. After killing his victim and walking up to the body, the Bedouins reported, he had taken off his sandals and run barefoot toward a road. They followed the track to the marks of a car and four sets of booted feet, and then to an encampment of six members of the Camel Corps.

Next day, the six men – and a number of others – were marched barefoot across an area of sand that had been specially smoothed. This was repeated several times, and each time the Bedouin trackers identified the same man. The police had their evidence, but it was insufficient for a trial. Fortunately, Smith was able to show that the bullet had been fired from the same man's gun. It transpired that the postman had been having an affair with the soldier's sister, and had been murdered to avenge the family's honour.

CASE CLOSED

found both the overshoes and the sneakers, while his fingerprints matched those already recorded. Eventually, the two men confessed to over 50 local burglaries.

Apart from leaving impressed tracks in the surroundings, housebreakers may kick in a door, leaving a clear dusty print – especially if they are wearing rubber-soled shoes. They may also use their feet for leverage to move a heavy object such as a safe. All such traces are photographed as evidence.

In an interesting American case in Ohio, a car left the road, striking a bridge wall, and throwing a woman in the vehicle to her death. A highway patrol officer found an unconscious man in the back seat who, when he recovered, refused to name the driver of the car. The investigating officer took the man's shoes, together with the brake and accelerator pedals, and submitted them for laboratory examination. The man's right shoe had 18 striations on the left side, which exactly matched 18 striations on the brake pedal. There was also a puncture in the leather, its position and size matching a wire sticking out from the bottom of the pedal. The man was later convicted of vehicular homicide.

TIRE PRINTS

The tires of cars, whether parked or moving, can leave imprints as important as those of shoes. One can find several different types of tire prints at the scene of a crime. They may be impressions in earth, clay, mud or sand, snow, or other materials on the site; they may be in the form of prints transferred from a pool of blood, oil or spilt paint, or from a nearby patch of mud; or, in hit-and-run fatalities or traffic accidents, they may appear as clear bruise impressions on the flesh of the victim.

Tire tracks left by moving four-wheeled vehicles or bicycles are only usefully clear when the vehicle has travelled out of true, or turned a corner. Otherwise, when the vehicle is moving in a straight line, the print of the rear wheel is superimposed on that of the front wheel, resulting in a confused track that is difficult to analyze, and will almost certainly not be accepted as evidence in court.

Prints in blood, paint or dried mud, or on flesh, can be photographed, and easily compared with similar prints taken from suspect and sample tires. In the case of residual mud prints, it may be possible, if the mud is of the right consistency, to gauge the depth of the tread. This can only be done at the scene of the crime, and it is very difficult to preserve samples as evidence.

When the tracks are in the form of impressions, they can give a very accurate picture, not only of the tread, but of its condition. As these tracks are three-dimensional, photography cannot preserve all the details, and casts must therefore be taken. In moist soil and clay, the cast can be made with plaster of Paris, or one of a number of silicone rubber mixtures. A silicone spray is used before taking the cast, to prevent any soil adhering to the casting mixture. When the impression is in dry soil or sand, a preliminary cast is made with a solution of shellac or a similar plastic material, from which a plaster cast can subsequently be obtained. Tracks in snow are seldom clean-cut, and must be dusted with talcum powder before the shellac cast is made.

The construction of a modern motor tire is complex, but the principal identifying factor is the tread. Every manufacturer has its own distinctive patterns, claiming the best grip and water-shedding properties for a variety of uses. Each company has a number of different patterns for cars, trucks, motorcycles and bicycles, and reference to a "library" of tire prints can quickly identify the make and type.

British police discovered the body of Jini Cooppen in the early hours of March 31, 1990, in Brixton, south London. She had been strangled, but there was no sign of a struggle, and there were also indications that she had been killed elsewhere. Consequently, the police were particularly interested in the tire marks they found not far from the body.

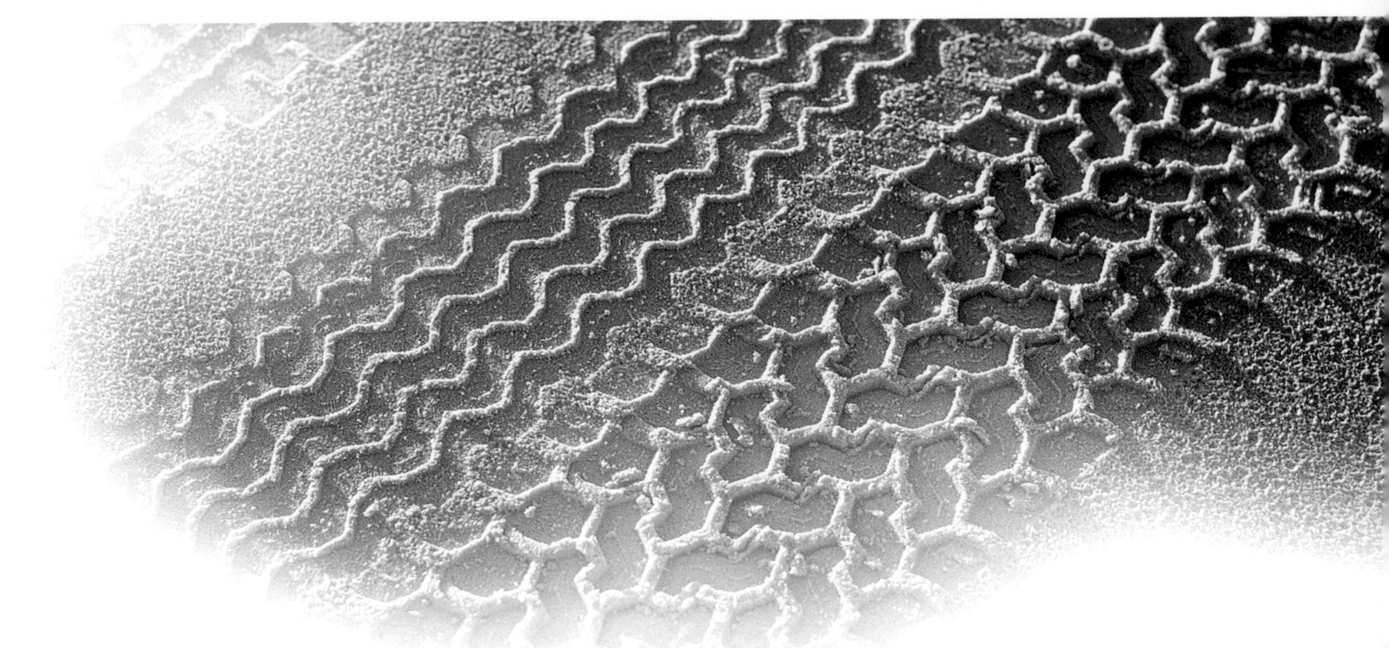

Suspicion centred on the victim's husband, Vijay Cooppen. The forensic report showed that the tire marks had been made by a vehicle with a Dunlop on the front nearside wheel, and a Goodyear on the front offside. Cooppen's Volvo had three Dunlop tires and one Goodyear, newly fitted on the morning of March 30. A comparison of the two front tires with the prints showed a perfect match, but this was insufficient to prove that they were identical. Therefore, the police decided to trace the batch of tires from which Cooppen's Goodyear had come.

There were only twelve moulds in existence that were used to manufacture the particular type and size of tire, and only two of these would have produced the relevant pattern. Most of the tires from these moulds were exported to Holland, and only a small percentage went to fitters in Britain.

The likelihood that another Volvo, carrying a Dunlop on its nearside front and the particular type of Goodyear on the offside front, could have been in London on March 30 was infinitesimal. The case was clinched when Cooppen's five-year-old son told the police that his father had been out of the house between midnight and 5 AM on the night in question.

A forensic scientist can gather a lot about the age of a tire and the vehicle carrying it from a flat print, and even more so from an impression. Tires come in different widths for different makes and types of car. The degree of wear, particularly in the case of an impression, can be gauged from the relation between the raised portions and the grooves. The loading of the vehicle can be determined from the width of the print. Unevenness in the car's suspension causes variations in wear from one side to the other; "scrub" due to improper alignment of the front wheels may also be apparent. Finally the forensic scientist looks in particular for any signs of specific damage that can be compared with a suspect tire.

Forensic laboratories maintain "libraries" of the different patterns of tire tread used by manufacturers, and their range of sizes. The tire's impression can provide information of the amount of wear, the loading of the vehicle, and even details of the suspension and alignment of the wheels.

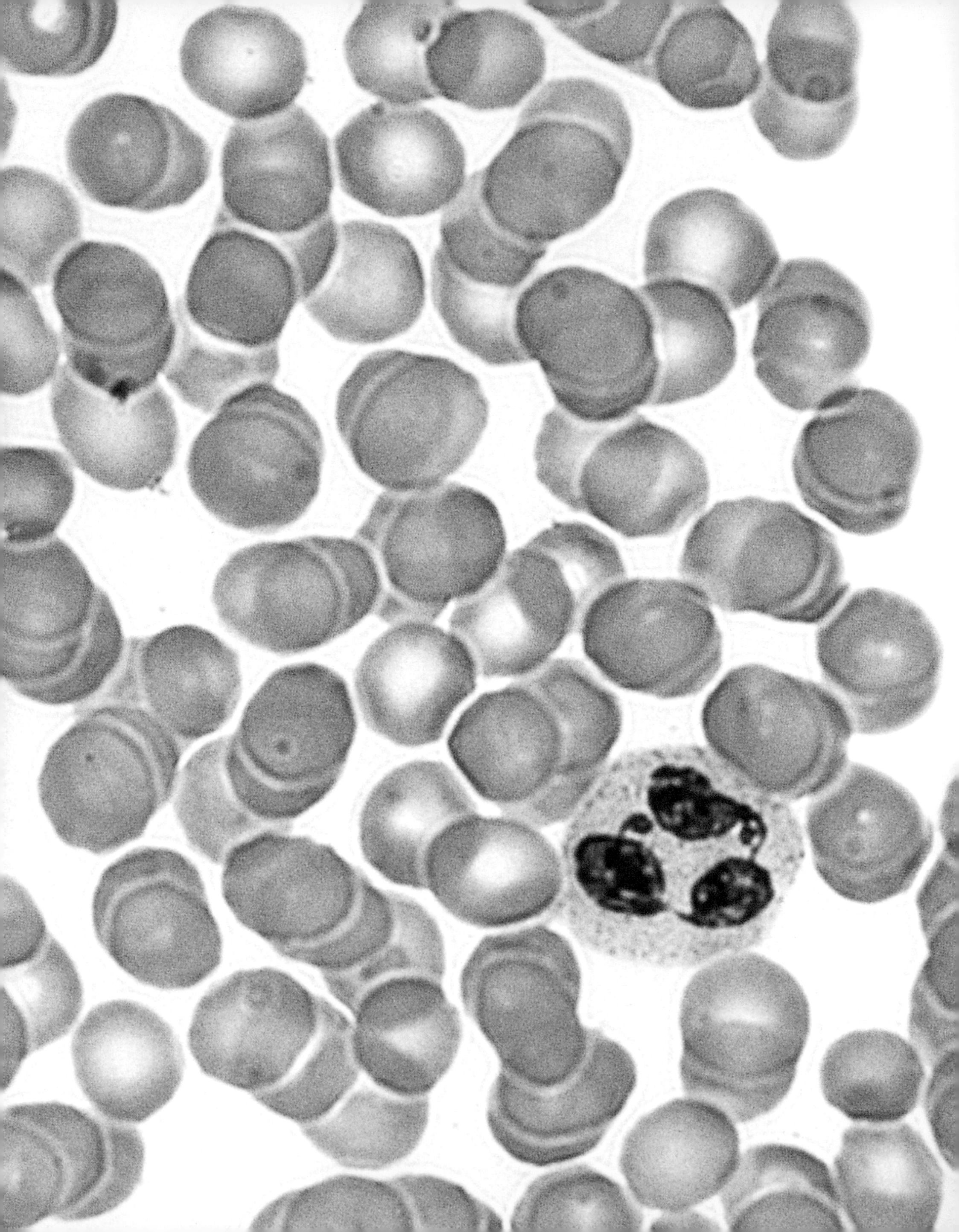

Written in Blood

THE AVERAGE HUMAN BODY CONTAINS 10 pints (6 litres) of blood. From earliest times, mankind has realized the connection between blood and life, but it was not until 1616 that William Harvey finally established the link between the beating of the heart and the continuous circulation of blood through the body. Soon after, physicians began to wonder whether it would be possible to replenish the blood supply, in a weak or dying person, by adding to it. However, except in a very few – extremely fortunate – cases, experiments in blood transfusion were doomed to failure. The reason for this was discovered at the end of the 19th century by the Austrian physiologist Karl Landsteiner.

Nearly 20 blood-grouping systems are now available to forensic investigators. Red blood cells, however, are of no use in DNA analysis, because they contain no nuclei. In this photograph, a single white blood cell has its nucleus stained purple. This is the only cell that can be used for DNA "fingerprinting".

The red blood cells carry substances called antigens, which help to produce antibodies to fight infection and disease. Landsteiner found that human blood could be divided into four main groups, depending upon the presence or absence of two specific antigens. He named the groups as follows:

Group A: antigen A present, antigen B absent.
Group B: antigen B present, antigen A absent.
Group O: both antigens A and B absent.
Group AB: both antigens A and B present.

All human (and primate) blood belongs to one of these four groups. The particular group depends upon genetic inheritance from parents, and the proportion of each can vary from one population to another. In Britain, the proportions are roughly: A, 42 percent; B, 8 percent; O, 47 percent; and AB, 3 percent. In the United States, they are: A, 39 percent; B, 13 percent; O, 43 percent; and AB, 5 percent.

Transfusion of blood from one person to another is only possible when both are of the same blood group. Mixing of blood from two different groups causes the red cells to clump together (agglutinate), something that is easily seen under the microscope. This happens because each blood group is associated with a related antibody, and incompatible antibodies cause agglutination.

The English physician William Harvey, who finally established that blood circulates through the body, in 1616.

Blood Identification

Clearly, identification of a blood type can be of great practical importance in forensic science. For example, if blood of type A, the same as the victim's, is found on the clothing of a suspect whose blood is type O, there is a strong suspicion – but no more – that it has come from the victim. Making use of other group systems, the probability can be greatly increased.

As an example: if blood of type O occurs in 47 percent of the population, the substance haptoglobin-2 occurs in 36 percent of type O persons, and the enzyme PGM-2 occurs in 5 percent of these, the probability of a person having these three blood types is 47 x 36 x 5 in 1,000,000 = 8460 – that is, approximately 8 in every thousand.

Landsteiner took a quantity of blood, and separated the blood cells from the liquid – the serum – by centrifuge. Then he added red blood cells taken from different people. He found that two distinctively different events took place: either the serum accepted the cells, or it seemed to repel them, causing them to agglutinate. Landsteiner's experimental method has since developed into a widely used laboratory analytical technique known as serology.

Later research has identified a number of other groupings. Landsteiner began to inject human blood into other species of animal: in this way he discovered in 1927 two other types of antigen. One he classified into M, MN and N; the other was the P group. Working in the United States with Rhesus monkeys, Landsteiner and A.S.Wiener discovered the Rhesus factor in 1940. Other researchers have introduced more than a dozen additional group systems.

More recently, blood typing has concentrated more on the different enzymes and proteins, associated with the major blood groups, that perform specific biological activities in the body.

Serologists use electrophoresis to detect the presence of specific enzymes and proteins in a blood sample. A piece of cotton thread is soaked in the blood, and pressed into a thin gel coating a glass plate. A weak direct electric current is then applied across the plate. The various components move across the gel, the distance they move depending upon their molecular size. After a set time the components separate, and can be made visible by staining with specific reagents. The results appear as a pattern of bands on the gel, not unlike the bar-codes used to identify products in stores. A similar technique forms part of DNA typing (see "DNA Fingerprinting").

Although DNA typing is rapidly superseding blood-group typing of this kind, many crime laboratories still carry it out. A simple ABO test, for instance, may be sufficient to eliminate a number of suspects in a murder or assault case. It has also been widely used in questions of paternity. However, some substances in blood are destroyed by heat or drying, and others are light-sensitive.

An important discovery concerning blood groups was made in 1925. About 80 percent of humans are "secretors" – that is, their saliva, semen, urine, perspiration and other tissue fluids contain the same substances as their blood. Even if blood is not discovered at a crime scene, therefore, other evidence can help to identify the perpetrator.

In 1949, two British scientists discovered that it was possible to distinguish between male and female body cells, particularly the white blood cells and those of the lining of the mouth. In the nuclei of female cells there is a body – named the Barr body after one of its discoverers – that will stain darkly. The Barr body is absent in male cells.

When stains are found at a crime scene, or associated with a suspect, investigators must first ask: are these stains blood or not? There are a number of tests that can establish this. One is the Kastle-Meyer test, which depends upon the fact that blood contains the enzyme peroxidase. Some of the suspected stain is extracted with a damp filter paper, and treated with a mixture of phenolphthalein and hydrogen peroxide. A pink coloration indicates the presence of peroxidase. Another test used the chemical benzidine, which gave a blue coloration, but this has been discontinued, due to the dangerously carcinogenic properties of benzidine.

Austrian physiologist Karl Landsteiner, who first discovered that human blood could be divided into four basic types, and established the modern science of serology. Later, working in the United States, he discovered other types of blood group, including the Rhesus factor.

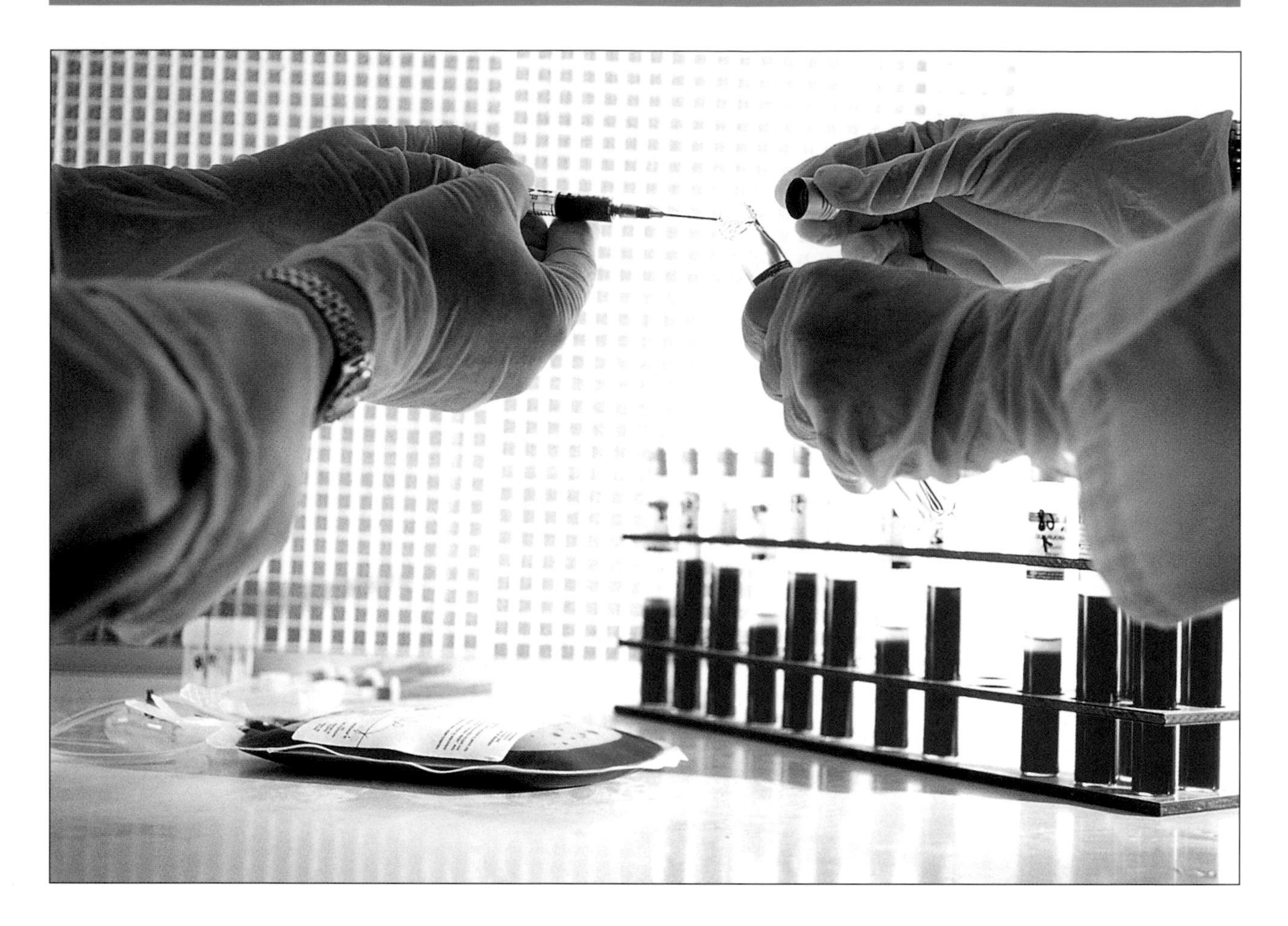

One stage in the examination of blood samples. Serologists can distinguish between human and animal blood, determine human sex, and, with a battery of tests, identify the blood as that of only a few possible people in a thousand.

The next question is: is this human blood, or that of some other animal? In 1901 a German biologist, Paul Uhlenhuth, came up with a test to answer this question. He discovered that, if he injected protein from a chicken's egg into a rabbit, and then mixed serum from the rabbit with egg white, the egg proteins separated from the clear liquid and formed a cloudy precipitate – precipitin. The rabbit's blood had manufactured antibodies to the egg protein, and produced a reaction similar to the agglutination of red blood cells. From this, it was an easy step to produce a range of serum specific for any number of animal blood samples.

In modern applications of the precipitin test, the blood in question is placed in one depression of a glass slide that has been coated with gel, and the specific serum is placed in an adjacent depression. A weak direct electric current passed through the gel causes the two samples to migrate toward one another, and a visible precipitin line forms between them. The test is very sensitive, and positive results have been obtained from dried blood that is more than 15 years old. Even tests on tissue samples from ancient mummies have proved successful.

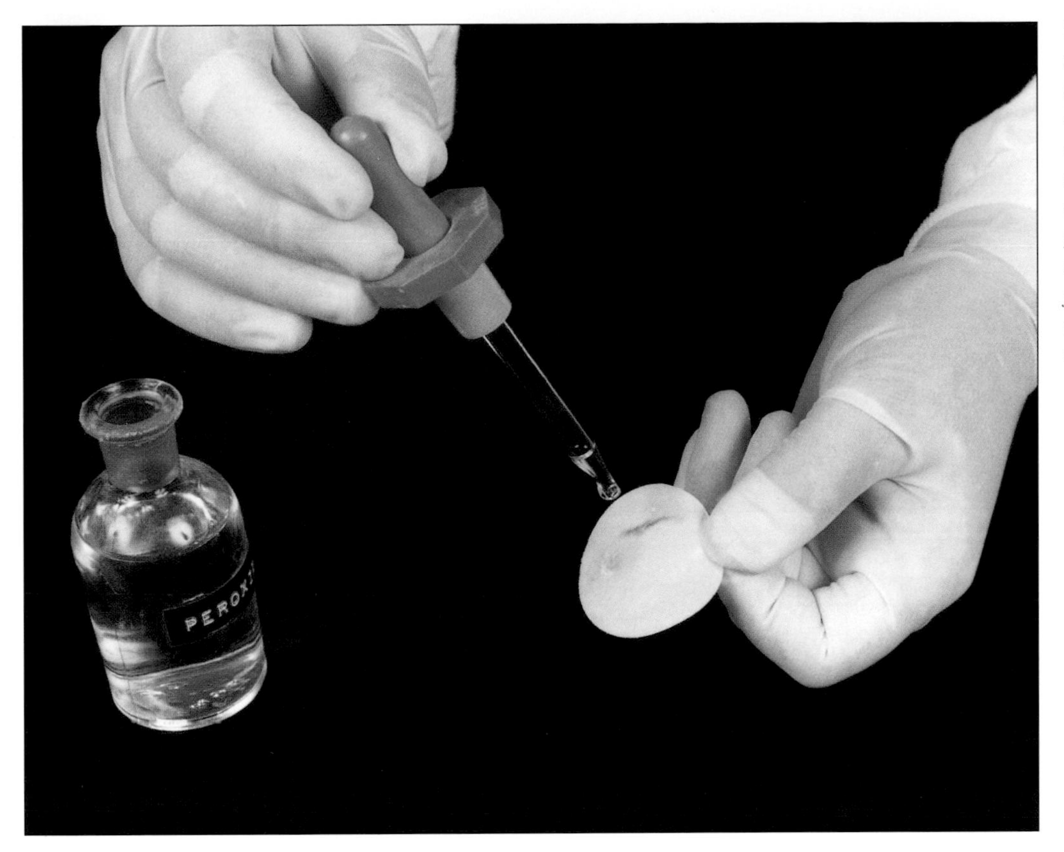

The Kastle-Meyer test is used to distinguish blood from other similar-looking matter. The analyst is about to apply a drop of a mixture of hydrogen peroxide and phenolphthalein to the filter paper. A pink coloration will indicate blood.

Serology has now developed to a point where completely automated equipment can submit a sample of blood to a battery of tests within a few minutes. One scientist pointed out that he could even distinguish between the blood of his twin daughters, because one had been infected with chickenpox, and the other had not.

With all these criteria at his or her disposal, the serologist can distinguish between human blood and that of other animals, determine human sex, and narrow down the owner of the blood to a few people in every thousand. On its own, this may be insufficient to convict a murderer, but it is powerful incriminating evidence.

TRACES OF BLOOD

Where a violent homicide has taken place there are often quantities of blood, not only on the victim, and any discarded weapon, but splashed on the surroundings, and even the ceiling of a room. The Scottish pathologist, Professor John Glaister, classified blood splashes into six types in the 1930s.

Drops are found on horizontal surfaces. They are essentially circular in shape, depending upon the height from which they have fallen. The greater the height, the more the impact will cause them to spray out into a star-like shape.

CRIME FILE:

Graham Backhouse

He believed he had successfully distracted attention from himself as the murderer of a neighbour, and the attempted murderer of his wife. But traces of blood at the scene of the crime told a very different story.

Graham Backhouse, walking with his two children.

Graham Backhouse gave up his business as a hairdresser when he inherited his father's farm in the village of Horton, in the English Cotswolds. But he was not a successful farmer and began to lose money. An inveterate womanizer – despite his wife of ten years and two young children – he outraged many of the inhabitants of the picturesque little village by his behaviour.

In the spring of 1984, Backhouse contacted the police and said that his life had been threatened. Soon after, his stockman found a sheep's head impaled on a fence, with a note that read "You next". On the morning of April 9, Backhouse offered his wife the use of his own car, As soon as she turned the ignition switch, it exploded in flames. Mrs. Backhouse was not killed, but suffered severe injuries. Explosive experts discovered that she had been the victim of a pipe bomb. It seemed there was substance in Backhouse's claim that his life was in danger, and a full-time police guard was mounted on the farm.

CASE CLOSED

Questioned by the police, Backhouse admitted that there were a number of people whose wives he had seduced, and who no doubt would be glad to see him dead. He also mentioned a near neighbour, Colyn Bedale-Taylor, with whom he had been in dispute over a right of way.

Blackhouse asked the police guard to leave just nine days after the bombing. An alarm was installed instead, linked to a police station some distance away. On the evening of April 30, the alarm sounded. When police arrived, they found Backhouse covered in blood, with deep slashes across his face and chest. Lying beside him was a shotgun, and at the foot of the stairs lay the body of Bedale-Taylor, shot in the chest at point-blank range. In his hand was a craft knife. Backhouse told the police that Bedale-Taylor had attacked him in the kitchen with the knife and, in self-defence, he had grabbed his shotgun and killed him.

The first indication that all was not as it seemed came when Geoffrey Robinson, from the Home Office Forensic Laboratory in Chepstow, examined the blood splashes at the scene. The drops were round, indicating that Backhouse had been standing still, or moving slowly, as he bled. If there had been a violent struggle such as he described, the blood would have been flung about, landing in a characteristic exclamation mark shape.

In addition, some kitchen chairs had fallen on top of the round spots, and

Opposite page (top): A forensic examiner displays the threatening notes that Backhouse alleged he had received. The "You Next" message was written on a page torn from a pad, and showed the impression of a doodle drawn on a previous page. When police searched Backhouse's office, they found the pad – and the doodle.

one had long smears of Backhouse's blood on its top – but there was no blood on his gun. Finally, although Bedale-Taylor's body lay at the far end of the passage leading to the kitchen, there was no trail of blood along it. Robinson suggested that Backhouse's wounds were self-inflicted while he stood in the kitchen.

On May 13, 1984, Backhouse was arrested, and charged with the murder of Bedale-Taylor and the attempted murder of his wife. In evidence, pathologist Dr. William Kennard stated that "the wounds could have been caused by someone else, but Backhouse would have had to stand still there doing nothing while his attacker slashed him from shoulder to hip".

As for motive, the court learned that Backhouse had accumulated bank debts of more than £70,000. In March 1984 he had increased the insurance on his wife's life to £100,000. He was found guilty and sentenced to two terms of life imprisonment.

Below:* *Police examine the wreckage of Mrs. Backhouse's car after the explosion.

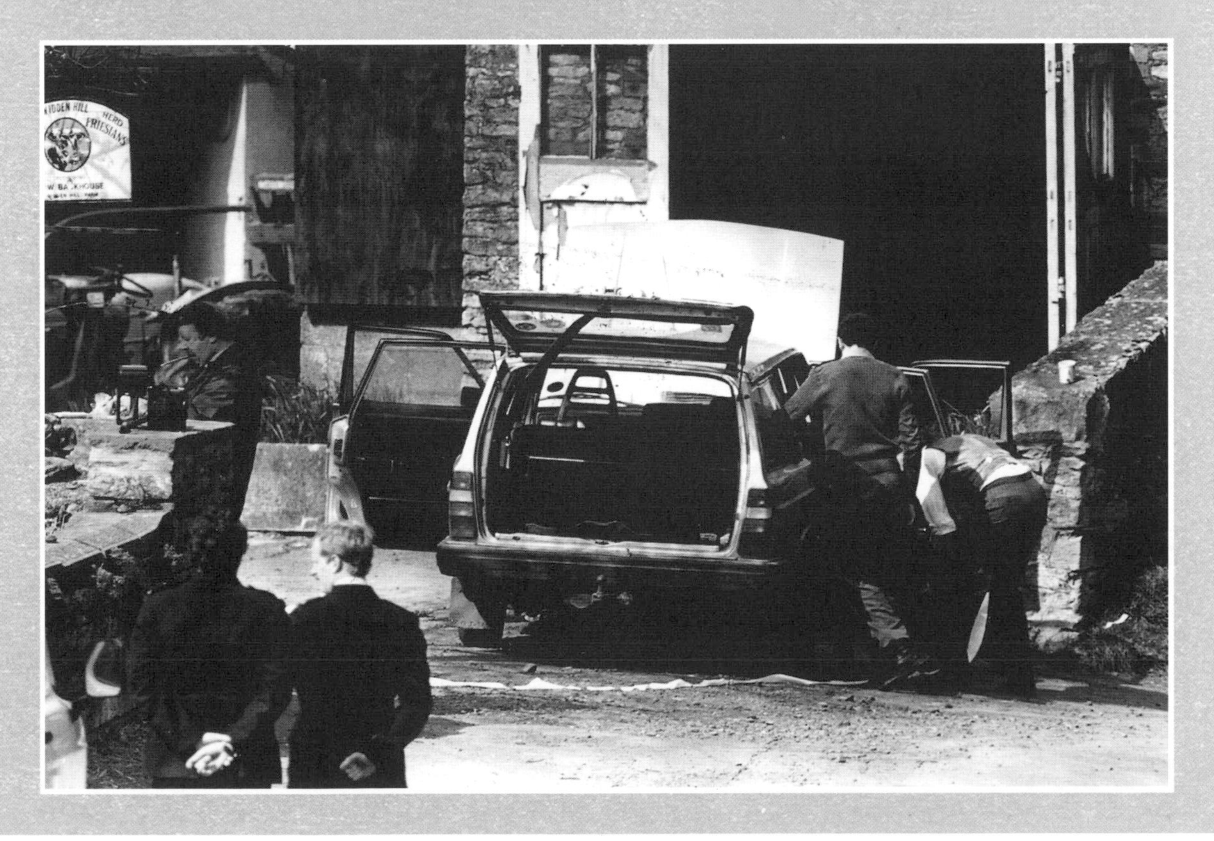

Part of the forensic serology laboratory at the Federal Bureau of Investigation at Quantico, near Washington , D.C.

Splashes occur where the blood has flown through the air, and hit a surface at an angle. This is most likely to happen when the victim has been struck with a moving weapon. The bloodstain is shaped like an exclamation mark, and its elongation indicates the direction of travel.

Spurts are due to the pumping action of the heart while the victim is still alive. If a major artery or vein is severed, the pressure can send the blood a considerable distance, reaching walls and ceiling of a room, and often staining the clothing of the assailant.

Pools form around the body of a bleeding victim. They can show where he had dragged himself from one place to another, or been dragged there.

Smears may be left on the surroundings by a dying victim as he attempts to move, or by the bloodstained assailant.

Trails indicate that a bloody corpse has been moved from one place to another. If the body is dragged, the trail will be smeared; if it is carried, there will be blood drops along the way.

Careful observation of the bloodstains at the scene of crime has proved invaluable in the investigation of murder cases, and alleged assaults.

CRIME FILE:

Ludwig Tessnow

A travelling carpenter was suspected of the horrific murder of two young boys in northern Germany. He claimed the stains on his clothing were dye, but a newly developed analytical technique revealed them to be human blood.

On July 1, 1900, two young brothers went missing from their home on the island of Rugen, off the Baltic coast of Germany. Next morning their disembowelled and dismembered remains were found scattered through a local woodland. A travelling carpenter, Ludwig Tessnow, was questioned, and dark stains were found on his boots and clothing. He claimed that they were wood dye, but the local examining magistrate remembered a newspaper report of a similar case in Osnabruck, three years previously and several hundred miles away. Two young girls had been brutally butchered in a similar manner; a man taken in for questioning had claimed that the stains on his clothing were wood dye – and his name was Ludwig Tessnow.

The magistrate discovered that, three weeks before the murder of the two brothers, a farmer had seen a man running from his meadow. When he investigated, he found that seven of his sheep had been hacked to pieces. Faced with Tessnow, he identified him as the man in question. Tessnow maintained his innocence, but news of Paul Uhlenhuth's research had become public. He was asked to analyze the stains on Tessnow's clothing, and in August 1901 he made his report. Many of the stains were human, and others were of sheep's blood. Tessnow – popularly dubbed the "Mad Carpenter" – was found guilty, and executed in 1904.

CASE CLOSED

The shape of a trace of spilt blood can tell a great deal about the conditions in which it fell. Drops (upper left) fall vertically on to horizontal surfaces, and their shape will be affected by the height from which they fell. These drops have fallen from 1, 3, and 6 ft (0.3, 0.9 and 1.8m) respectively. Splashes (lower left) fly through the air, and hit the surface at an angle. The characteristic "exclamation mark" shape indicates the direction in which they have travelled.

DNA Fingerprinting

SINCE 1984, WHAT PROMISES TO BE THE MOST INCONTROVERTIBLE technique for the identification of an individual has been steadily gaining ground. It is popularly known as DNA fingerprinting. Despite the somewhat exaggerated claims that were made for it during its early development, it has already established its importance, not only in criminal investigation, but in questions of paternity and genealogy – and even in the evolution of prehistoric animals. Furthermore, whereas identification by fingerprint requires a substantially identifiable portion of a complete print, DNA analysis requires only a few individual body cells.

The nucleus of every human cell contains 46 paired chromosomes, and each comprises a number of genes. The chromosomes are nucleoproteins, made up of protein and DNA. In this micrograph, the chromosomes are stained deep red, and fragments of DNA fluoresce a yellow-green.

The initials DNA stand for deoxyribonucleic acid. This substance makes up the genetic material of all the cells of the body that contain a nucleus – tissues, bone marrow, hair roots, tooth pulp, semen, and white blood cells, and also waste cells in saliva and urine – but not the red cells of the blood, which have no nucleus. The story of the discovery of DNA, and the unravelling of its molecular structure, is a fascinating one, but there is no need to detail it here. The description of its structure and function that follows will therefore be necessarily somewhat simplified.

The DNA molecule can be visualized as a long ladder twisted into a tight helix. The two sides of the ladder are made up of alternating groups of phosphate (P) and the sugar deoxyribose (S). The rungs of the ladder are formed by groups of two "purine bases" linked at each side to a sugar molecule. There are four bases – adenine (A), thymine (T), guanine (G), and cytosine (C). Each rung contains two of these; but adenine can only be joined to thymine, and guanine to cytosine. So a single rung can be S-A-T-S, S-T-A-S, S-G-C-S, or S-C-G-S. Human DNA contains about three billion of these rungs.

When a cell divides, the two halves of the ladder separate, and each half then acts as a model for the formation of a new DNA molecule. Each is made up of a succession of phosphate, sugar and one base: this unit is called a nucleotide. A gene consists of a group of nucleotides, and provides the code for the formation of specific amino acids and enzymes. Each amino acid or enzyme governs a particular aspect of the body's metabolism, and determines the inherited physical characteristics, such as eye and skin colour, blood type, and so on. Since no two people – except twins who have developed from a single egg – are completely identical, they differ in a number of their genes.

The Human Genome Project, to which researchers all over the world have contributed, is devoted to the identification of each one of these genes.

Many of the gene pairs (a succession of ladder rungs) that are inherited are identical: for instance, most humans are born with two eyes, two ears, two legs and arms, etc. These are called class characteristics, and distinguish one species from another. When two parents provide the same gene, the result is what we commonly call "family likeness". If the two parents' genes are different, one may be "dominant". For instance, a child who inherits two blue-eye genes is blue-eyed; but if he or she inherits one blue-eye and one brown-eye gene, the eyes are brown, because the gene for brown eyes is dominant. Racial characteristics are inherited in the same way.

Restriction enzymes will cut the DNA molecule into fragments of different lengths. The "Southern blot" technique can then be used to separate the fragments into two separate strands. In this diagram, the DNA has been cut at two points by different restriction enzymes.

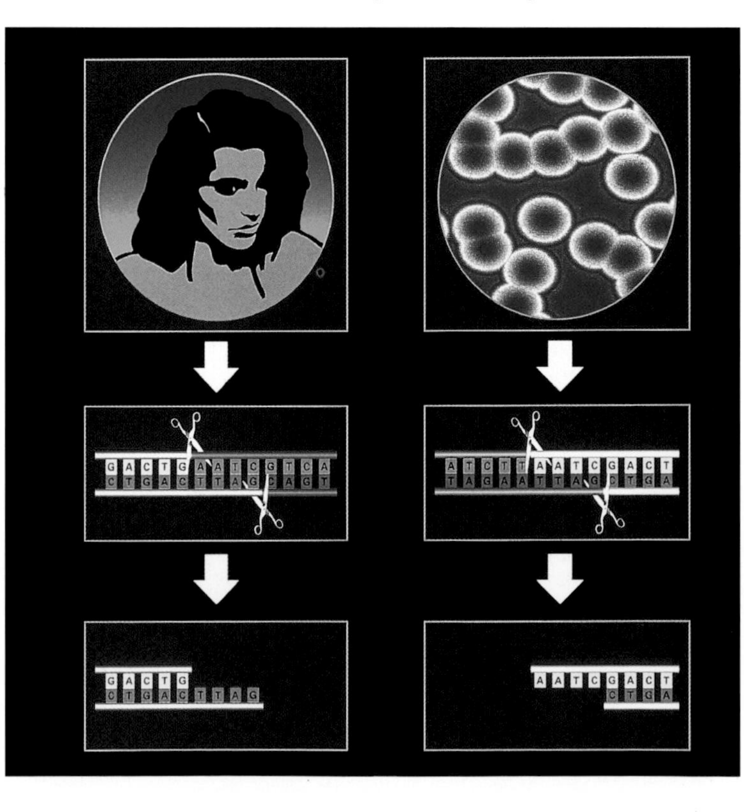

Enzymes are catalysts that promote the thousands of complex chemical processes that go on in the living body. Biologists have discovered certain enzymes that will cut up a strand of DNA into sections. These are called restriction enzymes. They are produced by bacteria to attack foreign DNA, and so protect themselves against viruses.

Since the sides of the DNA ladder are made up solely of alternating sugar and phosphate groups, we can identify a length by means of the base pairs that make up the rungs, without

CRIME FILE:

Lisa Peng

It looked like another case of violent rape, and a bite mark on the victim's arm seemed to support this belief. But traces of saliva from the bite provided sufficient DNA evidence to identify the true perpetrator of two murders.

Jim Peng, a Taiwanese living in Orange County, California, visited Jennifer Ji, the mother of his baby son Kevin, at her home on August 18, 1993. A shocking scene met his eyes. Ji's body lay against the sofa in a pool of dried blood. His son had been suffocated in his crib. When police arrived, Peng handed over a button that he had found on the floor. It appeared to have come loose from a woman's dress – but not Jennifer Ji's.

At first sight, this seemed to be a case of rape: Ji's underpants had been pulled down, and she had been stabbed at least 18 times. Vaginal swabs taken from Ji did not reveal semen, but this still did not rule out attempted rape. Support for the theory came from a round wound on her left arm: it was a bite mark, and swabs were taken from the wound.

Peng was a natural suspect, but he appeared to be innocent. Then the sheriff's investigators learnt that Peng's wife Lisa was visiting him from Taiwan. They paid the couple a visit, and Lisa Peng agreed that the investigators could search for a dress missing a button. They found nothing – except for two bags containing women's dresses, underwear, even shoes, all slashed to ribbons.

Embarrassed, Jim Peng explained. At some time in the previous year, his wife had arrived unexpectedly, and found Ji living with him. In a fury, she had destroyed the other woman's clothing. Faced with this evidence of Lisa Peng's violent temper, the investigators decided that they had a prime suspect for Ji's murder. An odontologist made a wax impression of Lisa Peng's teeth. It matched the photographs of the bite on Ji's arm, but was not in itself sufficient to justify an arrest.

Meanwhile, the swabs from the bite had been subjected to PCR analysis, which detected a locus common to 20 people in 100. Lisa Peng had returned to Taiwan, and a blood sample for comparison had not beeen taken before she left. However, one of the forensic experts remembered the wax impression, and hoped that sufficient saliva had been left on it. Sure enough, it too identified the same locus, but the probability of 20 in 100 was still insufficient.

A second PCR test identified a different locus. This, too, gave a match between the saliva from the bite and that from the wax – and it indicated a probability of only 1 in 200. Taking both probabilities together, this gave a figure of 20 x 1 in 100 x 200, or 1 in 1000.

The saliva from the wound had provided enough DNA for the analysts to suggest setting up the more accurate RFLP procedure. Peng was told that the only way his wife could be eliminated from the investigation was by providing a sample of her blood. Knowing nothing of the DNA tests, she agreed to return to Orange County. The first run of RFLP analysis took nearly a week to complete, but, confident that she had left no blood at the scene of the crime, she remained in the United States. The tests gave a match, and she was arrested on January 7, 1994, and charged with murder.

CASE CLOSED

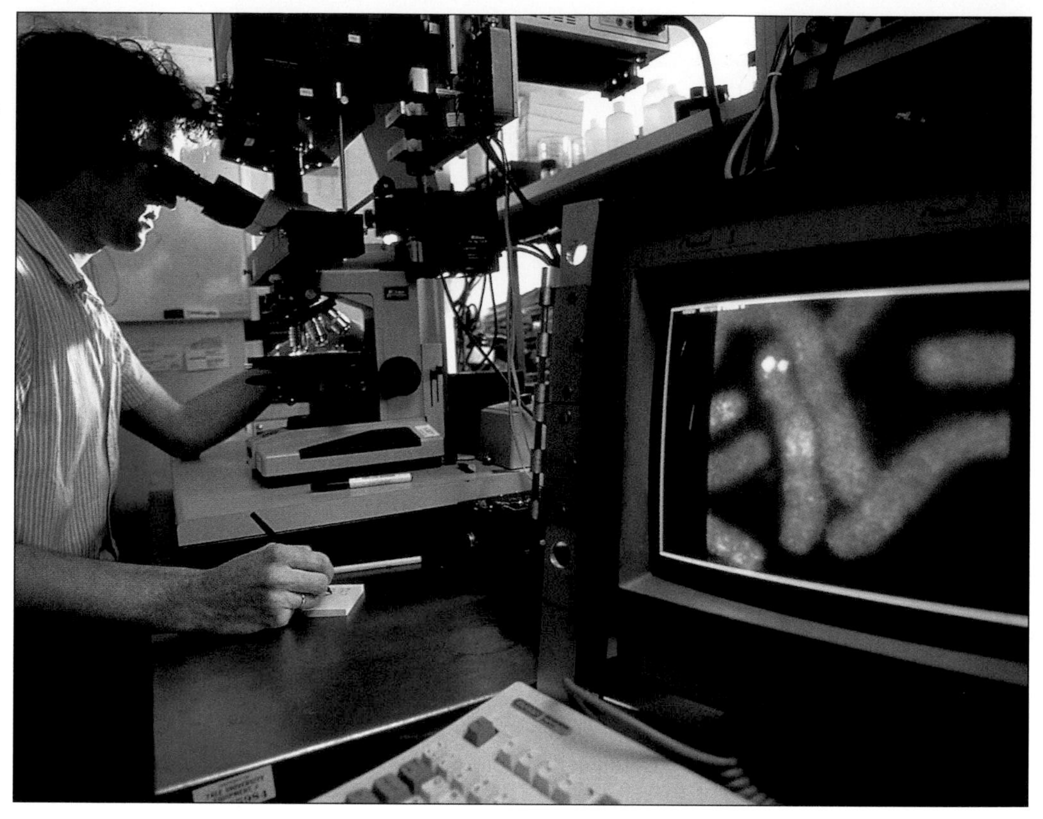

The mapping and sequencing of all the genes in human DNA has been the aim of the Human Genome Project, the completion of which has recently been announced. Here a researcher at Yale Medical School uses a projection microscope to study the gene structure of human chromosomes.

bothering to include -S- and -P-. For example, part of the sequence could be:

T-A-T-G-G-C-C-C-C-T-A-T-T-A-C-G-C-G-T-T-T-A-G-G-C-C-T-T-C-G-A-T-T-
A-T-A-C-C-G-G-G-G-A-T-A-A-T-G-C-G-C-A-A-A-T-C-C-G-G-A-A-G-C-T-A-A-

Restriction enzymes only cut the strand at a specific sequence of base pairs. The enzyme frequently used by American forensic laboratories is produced by bacteria named *Haemophilus aegyptius*, and known as Hae III. This cuts DNA only where it finds the sequence:

-G-G-C-C-
-C-C-G-G-

In the above example, therefore, it cuts the strand at two points, leaving the fragment:

-C-C-T-A-T-T-A-C-G-C-G-T-T-T-A-
-G-G-A-T-A-A-T-G-C-G-C-A-A-A-T-

Another restriction enzyme, Hae I, cuts DNA only where it encounters the sequence:

-A-G-G-C-A-
-T-C-C-G-T-

Over 400 different restriction enzymes have been isolated so far, each of which cuts the DNA at a different place.

CRIME FILE:

Dudley Friar

Nasal swabs taken from a paper handkerchief at a crime scene were to provide a vital forensic clue that would lead New York police to the apprehension of a multiple rapist and murderer.

Dogged forensic work uncovered the identity of the man who raped and strangled three women in New York within three weeks, but DNA typing clinched the case.

The body of Louise Kaplan, a 34-year-old fashion photographer, was discovered early on the morning of October 7, 1990 among trashcans, in an alley between two hotels in central Manhattan. There were clear signs of rape: scratches on Kaplan's thighs and around her vulva, and "a copious amount of semen", as the medical examiner described it. She had been strangled manually, and it was the third similar crime to have taken place in that area of Manhattan in 23 days.

Kaplan's car was found parked 50 yards (45.5 metres) away. There were signs that there had been a struggle inside. Marks on the rear seat showed that someone had been sitting there, probably waiting for her. Kaplan had escaped from the car, but had been caught in the alley, where she had fought bravely before being overcome.

The police had obtained DNA typing from the previous two rapes; it indicated that the same man was likely to be responsible for both. For some inexplicable reason the analysis of the semen from Kaplan's body was unsuccessful. However, among the large quantity of rubbish taken from the scene of her murder, laboratory technicians found a crumpled paper handkerchief, stained with nasal mucus. Under the microscope they saw large clumps of white blood cells, an ideal source for DNA typing; on analysis, it matched the semen from the first two murders.

A detailed police search of Kaplan's car uncovered on the carpet in the rear a clear dusty print of an unusually ribbed shoe sole. In the crevices of the upholstery of the rear seat, they discovered a number of dark blue fibres. Finally, there was a strange pattern of pinpricks in the padding of the roof.

Within a few days, a police burglary unit reported that the shoe print matched one that had been found at the scene of a liquor-store robbery two weeks earlier. Then a police officer said that he had seen something like the pinprick pattern before: it was the imprint of a cap-badge: "a showy kind of badge, like some security outfits issue to their employees". This suggested that the blue wool and polyester fibres had come from the uniform of a security guard. Within two days a fabric manufacturer in the Bronx produced a match, together with the names of the only three companies in New York to whom they sold it.

On the evidence of security video pictures from the liquor-store hold-up, the police had arrested three suspects. These consented to nasal swabs being taken, which were then sent for DNA testing. One was a match, and the laboratory stated that probably only six men in the whole United States would have precisely the same DNA profile. One of those was in custody.

His name was Dudley Friar: he worked at night as a hotel security guard and occasionally acted as a clown for children's parties. He was 29 years old, and had a 15-year record of minor theft and violence. The evidence against him was overwhelming; he confessed, and was sentenced to three terms of life imprisonment.

CASE CLOSED

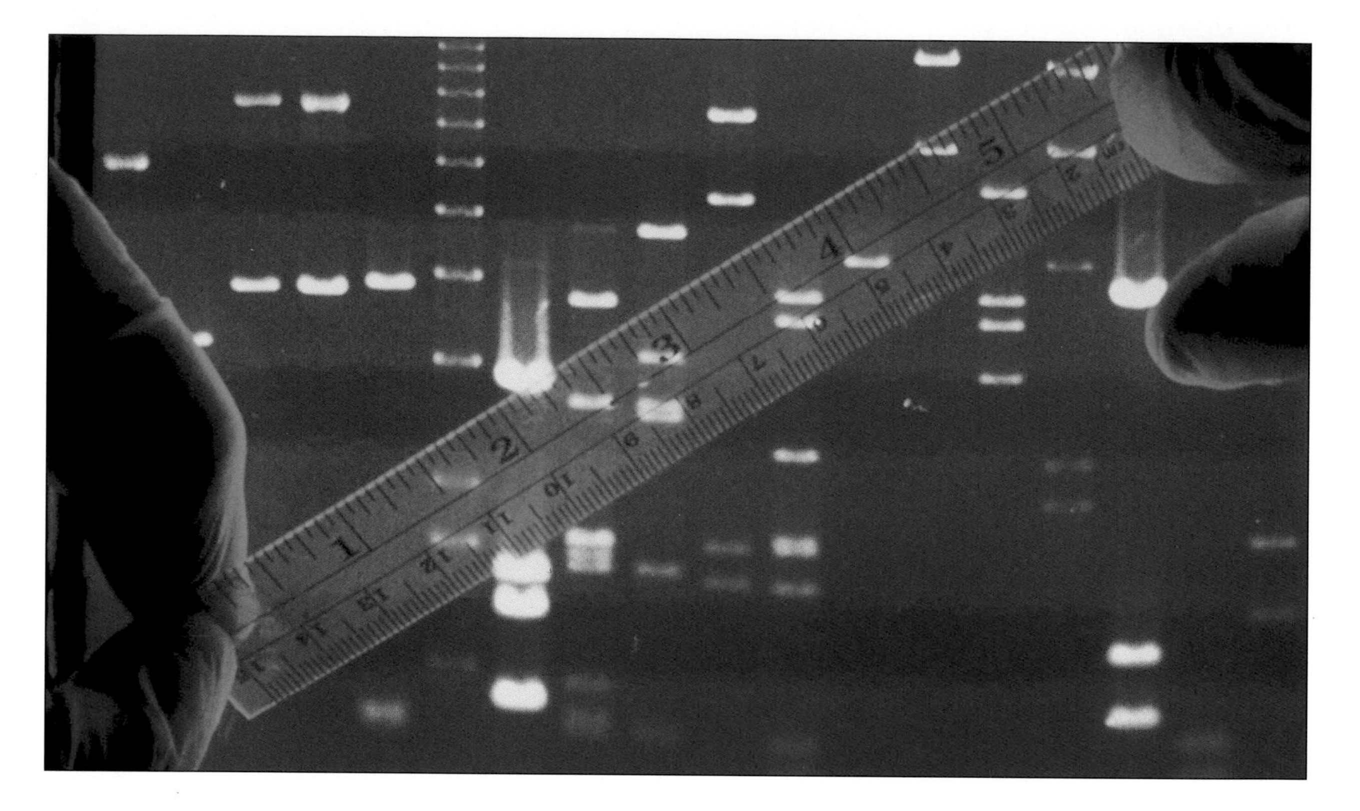

An autoradiograph of the analysis of DNA fragments by gel electrophoresis. Identical fragments, hybridized with a radioactive probe, have moved an identical distance across the gel.

ANALYZING THE FRAGMENTS

The different fragments of the ladder vary in length, according to the genetic structure of the individual's DNA. In other words, they have different molecular sizes. There is a well-established analytical procedure for separating molecules of this type according to their size: electrophoresis. The equipment is relatively simple: a glass plate coated with a thin layer of gel, across which a low-voltage direct current is applied. A DNA sample is placed at the negative end of the plate, and the molecules move across the gel at different speeds according to their size. Since the molecules migrate in a straight line from the negative pole to the positive pole, a number of samples can be placed side by side for comparison. After about 16 hours, the fragments have separated across the plate.

But how is the position of the individual fragments detected? The next step, developed by Edwin Southern in 1975, is known as the "Southern blot". The gel is soaked in a solution that separates the double-stranded fragments into single strands, while being pressed against a plastic membrane, on to which the single strands are transferred.

Opposite page: Single strands of DNA fragments. In this micrograph, a radioactive probe has located its complementary base.

The nucleotide bases of the strand – A, T, C, G – are now exposed along their length, and are treated with one or more "probes". A probe is a short length of single-strand DNA obtained by other means, and "labelled" with a radioactive atom.

CRIME FILE:

Carla

The rape and murder of a 12-year-old Bavarian schoolgirl in 1998 caused outrage throughout Germany. DNA samples once more provided the vital forensic evidence.

Twelve-year-old schoolgirl Carla was found in a pond near the Bavarian village of Frankendorf in January 1998. She was still alive, but in a coma, and died of serious injuries five days later without recovering consciousness. Investigators said Carla had been ambushed and sexually assaulted while on her way to school. When she tried to resist, she was strangled until unconscious and then thrown into the pond to make sure she was dead.

Police published a photofit of a man believed to have been in the vicinity at the time. The trail led them to an unnamed 31-year-old window installer, who denied the murder but admitted having been in the area at the time, an admission he later retracted. The killing, which happened shortly after another man was sentenced for an unrelated murder of a young girl, caused a nationwide uproar.

The most important forensic evidence took the form of a number of cigarette ends that were found floating in the pond. DNA in the saliva on the cigarette ends was found to match samples taken from Carla's body, and presiding judge Adolf Kölbl said there was no doubt that the accused had been the killer. In March 2000, the man stood staring impassively at the ground as he was sentenced to life imprisonment with a recommendation that he serve at least 15 years.

CASE CLOSED

The bases in the probe find their complementary bases in the fragments, and attach ("hybridize") to them. After washing, the membrane is placed in contact with a sheet of X-ray film, and the radioactivity of the probe produces an image on the film. The result is an "autoradiograph" of short dark bands across the film, not unlike a bar-code, each band representing a fragment. Bands that are level with one another indicate that the fragments are of the same molecular structure.

Mixtures of fragments of different lengths are known as "restriction fragment length polymorphisms" (RFLP) and RFLP analysis is the standard process for identifying DNA. In the United States, the use of probes that locate only one pattern of bases (a "locus") is preferred, because the results are more easily interpreted. This method is properly called "DNA typing".

CRIME FILE:

Tommie Lee Andrews

Despite an apparently unassailable alibi, DNA analysis conclusively proved that he was a serial rapist, in the first case in America in which such evidence was accepted.

The first American criminal actually to be convicted on DNA evidence was Tommie Lee Andrews, of Orlando, Florida. Between May and December 1986, 23 rapes at knife-point were attributed to him, and he continued his assaults during the early months of 1987. On February 22, fingerprints were found on a window screen, and on March 1 Andrews was arrested after a car chase. Although there was a considerable quantity of persuasive evidence – the first victim, Nancy Hodge, identified Andrews, his prints matched those from the February assault, and his blood type was the same as that in semen samples taken from victims – the police wanted more. Charged with the rape of Nancy Hodge, Andrews produced an apparently unshakeable alibi: he had been at home with his girlfriend and sister on the night in question.

The Assistant State's Attorney decided to obtain a DNA typing of Andrews's blood and the semen. They matched. After a pre-trial hearing, the judge agreed that the evidence was admissible; but the prosecutor made such sweeping claims about the probability of identification that the defence challenged the figures, the jury was hung, and a mistrial was declared. At a second trial for the rape of February 22, both the fingerprint and DNA evidence were admitted, and Andrews was sentenced to 22 years' imprisonment in November 1987.

In the retrial of the Hodge case in February 1988, the prosecution had properly prepared its case this time, and expert witnesses carefully explained the DNA typing technique to the court. Andrews was found guilty, being sentenced to jail terms totalling 115 years.

CASE CLOSED

Another method of RFLP analysis detects the succession of a number of identical base sequences – such as -T-T-A-T-T-T-A-T-T-T-A-T- – along the strand. These are called "variable number tandem repeats" (VNTR) because the number of repeated sequences can vary. It was a development of VNTR analysis that was given the name of "DNA fingerprinting" by the English scientist Alex Jeffreys in 1984. The procedure is more complex than DNA typing, and takes longer, because it uses probes that can recognize multiple loci. It is, however, much more discriminating than single-locus methods. Moreover, because it is claimed that no two individuals' DNA is identical, it merits the name of "fingerprinting".

A principal drawback of the RFLP technique is that the sample available for analysis may be very small. A technique known as the polymerase chain reaction, or PCR, has been recently introduced. It uses an enzyme that copies a DNA strand; this results in two strands, each of which is copied, making four, and so on in a chain reaction, so that a million or more copies can be made in a short time. In theory, at least, DNA from just a single cell nucleus can be replicated in this way. PCR kits are available to many law enforcement agencies in the United States, and can be used by relatively inexperienced technicians. Unfortunately, the test has its limitations: because each enzyme targets only a single locus, it can discriminate between comparatively few people.

Identifying the Individual

Assume that investigators have obtained a sample for DNA analysis – semen, white blood cells, a rooted hair – from the scene of a crime, and that another sample (which does not have to be of the same material, and is often a scraping of loose epithelial cells from inside the mouth) is obtained from a suspect. The autoradiographs match. Is this proof that the suspect committed the crime, or was at least present at the scene?

Forensic evidence of this kind has to be treated with caution, and it has taken time for it to be accepted by the courts. Many sweeping claims have been made for the specificity of DNA typing, and the prosecution must be very careful in arguing the likelihood that the two samples come from the same source. In more than one case in the United States, evidence has been disallowed when prosecuting counsel revealed insufficient understanding of the statistical mathematics involved.

When RFLP analysis is used, the probability (that a sample taken from a crime scene matches that taken from a suspect) has to be based on the frequency with which each identified fragment occurs within the population at large. Suppose, for example, that four fragments have been identified, with respective frequencies of 42, 32, 2 and 1 in 100. The probability that all four will appear in a single sample is 42 x 32 x 2 x 1 = 2688 in 100 million, or approximately 1 in 37,000.

Clearly, this requires detailed population statistics: these are only gradually being compiled. The calculations can be complicated by the fact that most crimes occur within narrowly-defined communities, often of very similar racial background. Unless the detailed frequencies for the local population – rather than the averages over the whole nation – are known, a claim that a suspect has been identified by DNA analysis may not be accepted by the courts.

Since 1998, local, state, and Federal agencies in the United States have established databases, much like fingerprint records. Crimes may then be solved by matching

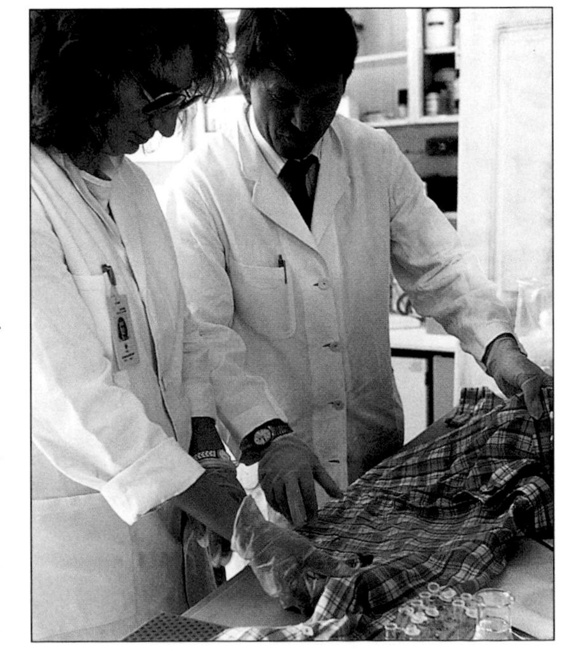

Bloodstained clothing from a crime scene is examined carefully by serologists. If blood typing shows that more than one type of blood is present, the "foreign" blood will be subjected to DNA analysis, so that it can be further matched to the DNA of a suspect.

CRIME FILE:

Colin Pitchfork

In one of the first DNA cases in England, analysis established the innocence of the youth charged with two rape murders, and a mass sampling campaign eventually led to the identification of the real culprit.

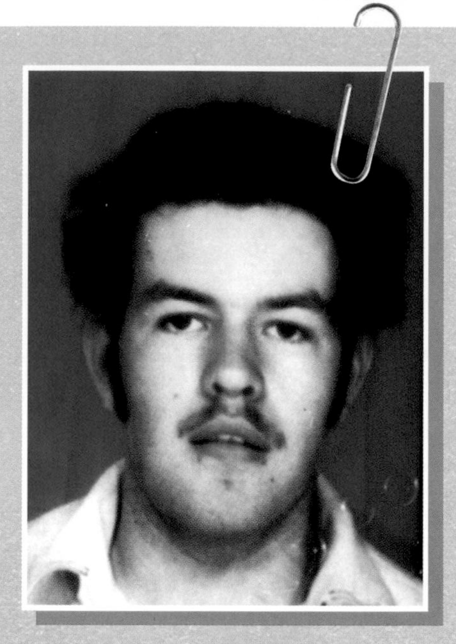

Colin Pitchfork's passport photograph. Pitchfork was one of the first people in Britain to be sentenced to jail on the basis of DNA evidence.

In one of the earliest cases of rape and murder where DNA profiling was used in England, it proved the innocence of one man who had confessed, and revealed the guilt of another.

On the evening of November 21, 1983, after visiting friends nearby, 15-year-old Lynda Mann failed to return to her family home in the village of Narborough in Leicestershire. At first light the following morning, a hospital porter, on his way to work at an adjacent psychiatric hospital, saw a body lying face-up in the grass. When the police arrived, they found Mann dead, stripped of her jeans and underpants, which lay in a pile nearby, and strangled with her own scarf. Dirt marks on her heels suggested that she had been dragged along the ground during the assault.

Mann had been raped before she was killed. Semen samples taken from the body established that the killer was a secretor of blood type A/PGM1+ (see "Written in Blood") – a classification possessed by about 10 percent of the adult male population of Britain. This figure was somewhat reduced by the high sperm content, which suggested a young man, and the police decided to narrow their search to males between the ages of 13 and 34. Despite inquiries by a 150-strong murder squad, and a computer search among all known local sexual offenders, no firm suspect was discovered. The hunt was finally called off in August 1984.

Two years later, on the evening of July 31, 1986, Dawn Ashworth, also 15 years old, disappeared while returning home from Narborough. Her body was found two days later, hidden in the grass not far from the same hospital. She also had been strangled, and sexually assaulted. The similarities convinced the police that the same man was responsible for this second attack.

At the time of their inquiries into Lynda Mann's death, the police had questioned a 14-year-old youth named Richard Buckland. He was notorious locally: big for his age, and not very intelligent, he sometimes startled women and girls by jumping out from hiding at them. However, he had been eliminated as a likely suspect. In the summer of 1986 he was working as a porter in the psychiatric hospital, and at dawn on August 8 he was arrested and taken to the local police station for questioning. After two days of

CASE CLOSED

rambling admissions, full of inconsistencies, he signed a confession and was charged with the murder of Dawn Ashworth – although he insisted that he was not responsible for the death of Lynda Mann. His trial was scheduled for November 21.

Because they believed that both crimes had been committed by the same man, the police asked Alex Jeffreys to carry out his DNA technique on a sample of blood taken from Buckland, an old semen sample from Lynda Mann, and a sample from Dawn Ashworth. After several weeks, he reported that one man had undoubtedly been responsible for both rapes – and that it certainly was not Buckland, who became the first murder suspect in legal history to be exonerated by DNA profiling.

This was a setback for the police, but they decided to take advantage of the new technique, and institute a voluntary mass sampling of blood and saliva among the young males of villages in the area. The sample of anybody found to be an A/PGM1+ secretor was then sent to the Home Office forensic laboratory at Aldermaston for DNA profiling. From January until September 1987, more than 5000 local males were tested, without success, and the police were under growing pressure from the Home Office to close down the inquiry

When the case broke, it was by accident. Chatting in a pub, a bakery worker mentioned that one of his colleagues, 27-year-old Colin Pitchfork, had paid him to give a blood sample in his name. The police computer revealed that Pitchfork was a convicted "flasher", and a former out-patient at the psychiatric hospital. He had been questioned at the time of the murder of Lynda Mann, but had arrived in the area only after the date of the crime. Without the availability of DNA profiling at that time, the police had no evidence to go on. Now, on September 19, 1987 they arrested Pitchfork, and sent a sample of his blood to Alex Jeffreys; the DNA matched. Colin Pitchfork was charged with the murder and rape of both Lynda Mann and Dawn Ashworth.

In a trial that lasted only a single day, Colin Pitchfork was sentenced to two terms of life imprisonment, and 10 years for each rape.

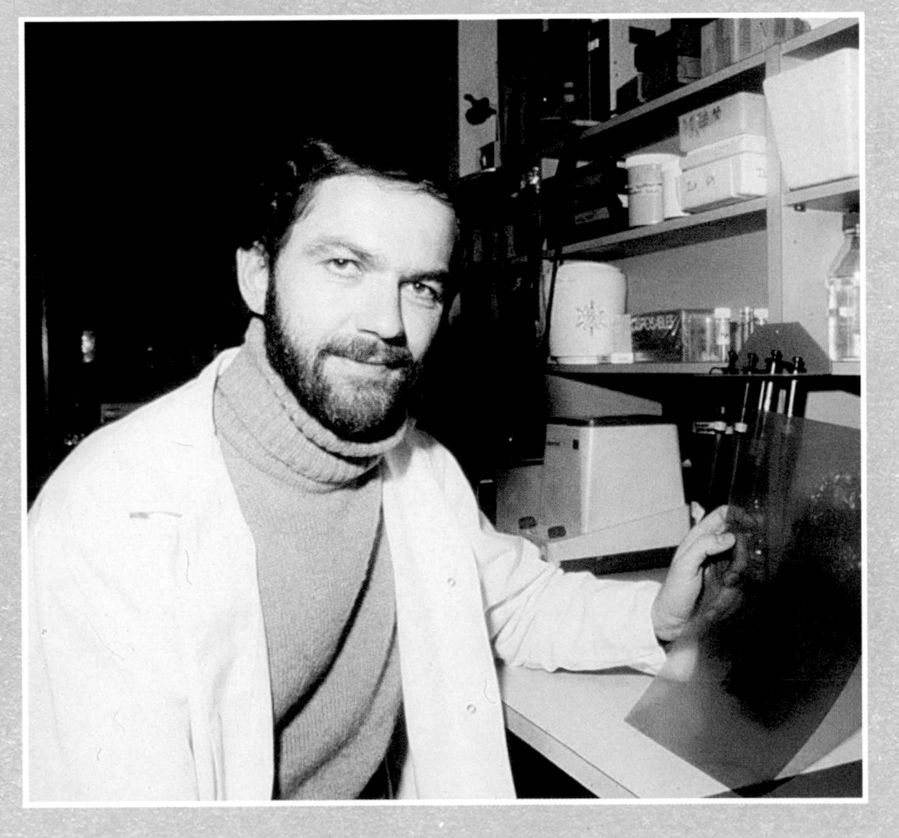

The English scientist Professor Alex Jeffreys, who developed DNA "fingerprinting" in 1984, holding a typical DNA scan.

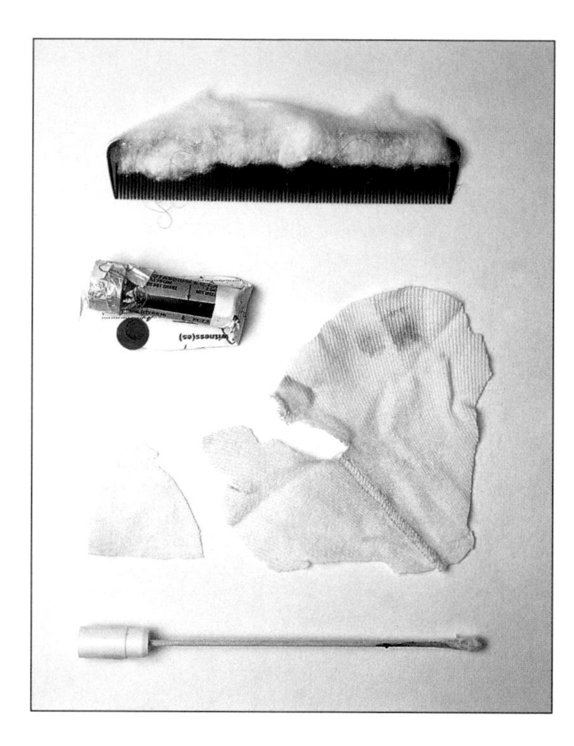

Typical samples from which a DNA fingerprint can be obtained in order to identify and convict a rapist, clockwise from top: pubic hair trapped in a comb with cotton wool; blood and semen stains on a torn pair of women's underwear; a vaginal swab containing semen; a semen stain on a piece of cloth; and a blood sample.

DNA typing results with those obtained from convicted criminals. The Federal DNA Identification Act of 1994 limits the database to DNA from convicted criminals. Access is restricted to law enforcement organizations, and a court order is required to use the information in judicial proceedings.

Every state collects DNA data of sex offenders. Individual states vary in whether they collect the DNA of other criminals, including murderers, robbers and those who commit crimes against children. White collar criminals are excluded. According to the Justice Department, in an experiment in 17 states over 3 years, 193 convicted criminals were matched to DNA taken from crime scenes.

In Britain, the Jeffreys method of DNA "fingerprinting" is the rule. Alex Jeffreys himself has claimed that, with his technique, the chance that two unrelated people have the same pattern is less than 1 part in 1,000,000,000,000,000,000,000,000,000,000 – a figure billions of times greater than the present world population! The earliest use of the Jeffreys technique was in paternity cases; the first application in a criminal case came in November 1987. A burglar broke into a house outside Bristol in southwest England in June of that year, raped a 45-year-old disabled woman, and stole her jewellery. Subsequently, a man named Robert Melias was arrested for burglary, and the rape victim was asked to attend an identification parade: she asserted that he was her attacker. The "bar-code" of the DNA from semen on the woman's clothing matched that of Melias's white blood cells, and he was tried and found guilty of rape and robbery. Further successes soon followed.

Since then, DNA typing and fingerprinting have been used increasingly in forensic investigations, and in other fields such as palaeontology. An amusing case occurred in England in the spring of 1997. Archeologists had recovered the skull of a prehistoric man from a cave in Somerset, and wanted to discover whether the DNA from the bone's interior indicated any kinship with that of present-day local inhabitants. Schoolchildren from the area were asked to cooperate by providing blood samples. None matched, but their teacher, who had taken part just to make up the numbers, was amazed to discover that he was undoubtedly the descendant of the man in the cave.

Now, British police have announced that the Human Genome Project is already so far advanced that they expect soon to be able to publish a full description of a wanted criminal, based solely on the genetic makeup of his DNA obtained from the scene of the crime.

CRIME FILE:

Jack Unterweger

A convicted killer, released as rehabilitated, he renewed his murderous career. DNA analysis helped to identify him, and the trail led through three European countries to Los Angeles, and back again.

Jack Unterweger, the serial killer, murdered eleven women, in three European countries and the United States, in two years.

In 1994, Manfred Hochmeister, a scientist at the Institut für Rechtsmedizin in Berne, Switzerland, was asked to carry out a DNA analysis on a hair found on a car seat associated with the murder of a Czech prostitute. As he had only a single hair root, he employed the PCR technique. This was the beginning of a chain of evidence leading to the conviction of an international serial killer.

The Czech police established that the car had been driven by Jack Unterweger. He had served a prison term in his native Germany for murder, and had taken up writing. He published a book and several plays, becoming quite famous, and on his release he turned to journalism.

Austrian police, meanwhile, were investigating several murders of women, whose bodies had been found in woods around Vienna. They soon realized that these had all been committed by the same person, and circumstances connected them with similar murders in Czechoslovakia and Switzerland. In every case, Unterweger had been in the area – and had even interviewed the chief of Vienna's police about the local murders.

Following the discovery of the hair, a warrant was obtained to search Unterweger's apartment. Police found a menu from a restaurant in Malibu, California, and photographs of the journalist posing with women members of the Los Angeles Police Department (LAPD).

Ernst Geiger, who was in charge of the Austrian investigation, immediately contacted the LAPD. They found that Unterweger had spent a month in Los Angeles in July 1991, staying in downmarket hotels, and claiming to be writing an article on prostitution for a leading German magazine. During that time, three prostitutes had been found on open ground, strangled with their bras. In each event, they had last been seen alive near a hotel where Unterweger had stayed.

The Austrian police called on the FBI's VICAP database (see "The Guilty Party"). Agent Greg McCrary, entering details of the European murders, found only four matches out of some four thousand cases – three of them the prostitutes strangled with their bras in Los Angeles.

At the trial of Unterweger, the court was told that, in little more than two years since his release from prison for a very similar murder, he had killed eleven women, in three European countries and the United States. He was found guilty of nine. A few hours later, he was found dead in his cell. He had strangled himself.

CASE CLOSED

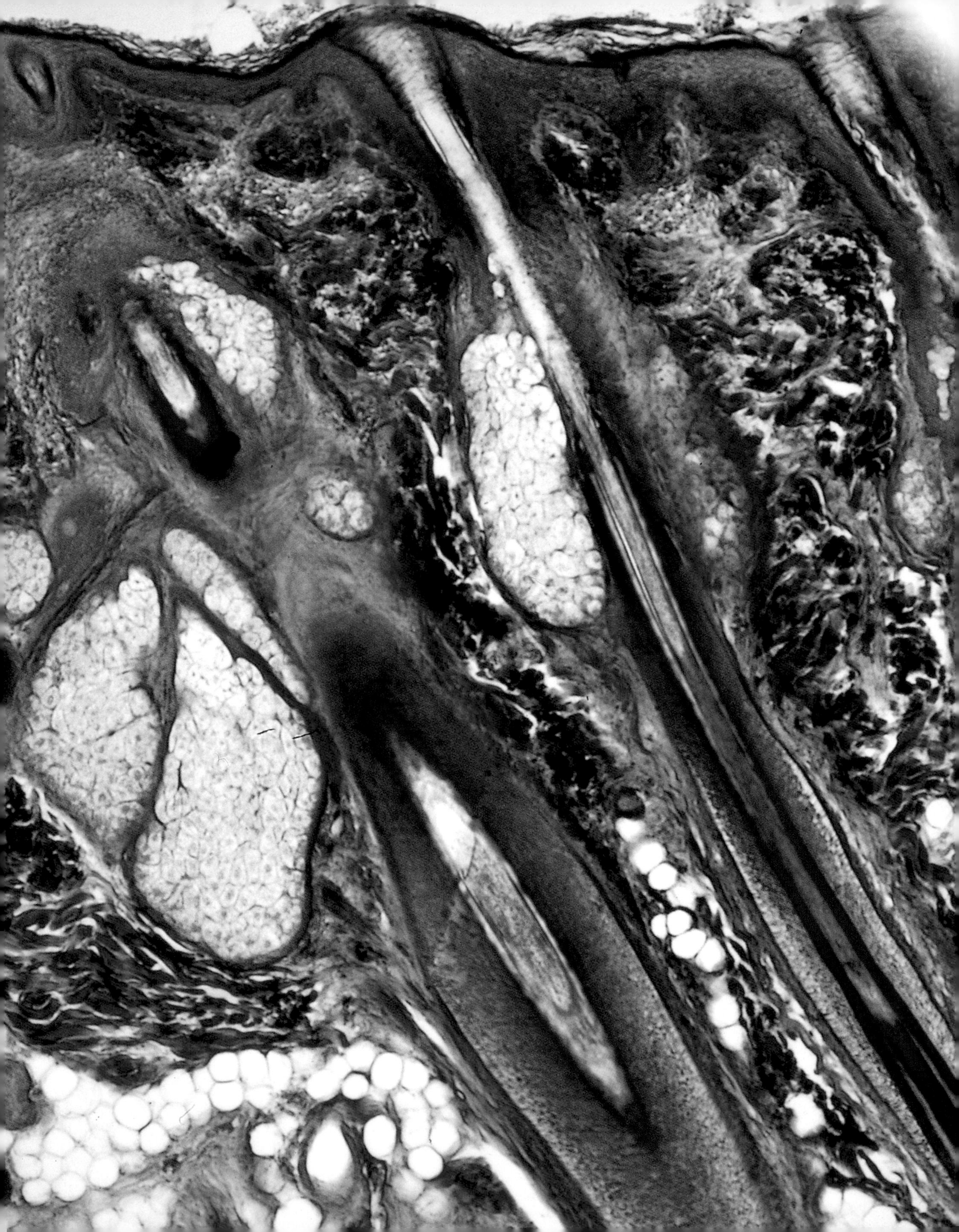

Hanging by a Hair

Hair grows from follicles in the skin. This micrograph shows three typical follicles, together with adjacent sebaceous glands. A hair bulb is clearly visible on the left.

Of all the possible contact traces left at the scene of a crime by a criminal, or carried away on his person, some of the most significant can be hairs and fibres. Human hairs have proved important in many cases, as they are relatively easy to distinguish. There may also be other hairs of animal origin, traces from woollen fabrics, silk, the mineral fibre asbestos, and a wide variety of plant or synthetic fibres, from clothing, carpets, cord or rope, sacking, or other materials.

The importance of hair in the investigation of crime has been recognized for many years: one of the first scientific papers on the subject was published in France as early as 1857. Professor John Glaister's 1931 work on the subject – *Hairs of Mammalia from the Medico-Legal Aspect* – remains to this day the standard reference.

Unless it is destroyed by fire, acid or alkali, hair remains identifiable on a body long after the flesh has decomposed, and it can also remain attached to a murder weapon. During life, the hair of the head grows on average about 0.09 inches (2.5 millimetres) per week, the male beard growing considerably faster, and the body hair more slowly. Growth ceases at death, but shrinkage of the skin, particularly in the face, makes it more

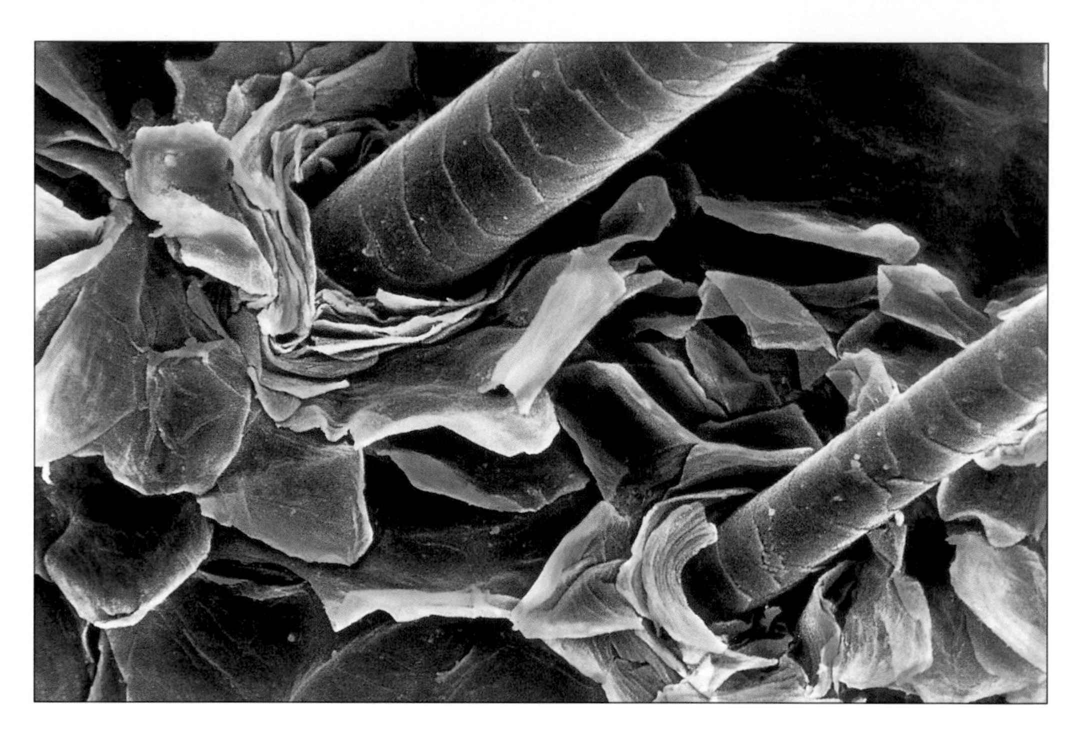

A micrograph of hair growing from a human scalp. The circular shape of the head hair is clearly visible, together with the characteristic epidermal scales.

prominent – which has given rise to the myth that beards continue to grow after death.

Hair is composed of protein, mostly keratin, and grows from follicles in the skin. Viewed longitudinally, a hair consists of three parts: the bulb, which is normally embedded in the follicle, the shaft, and the tip. In cross-section under the microscope it is also seen to have three parts: the cuticle

FACT FILE

Human hair is easily distinguished from that of any other animal, except for certain ape species. There are considerable variations in the structure of human hair, even in the same individual. Examiners divide it into six types:

- Head hair: generally circular in cross-section, with the ends – as the result of hairdressing – often split.
- Eyebrow or eyelash hair: circular in cross-section, but with tapered tips.
- Beard or moustache hair: stiffer, generally curlier than head hair, often triangular in cross-section.
- Body hair: oval or triangular in cross-section, generally curly.
- Axillary (underarm) hair: oval in cross-section.
- Pubic: usually springy, and oval or triangular in cross-section. Female pubic hair is generally shorter and more coarse.

The age of hair can be ascertained only approximately, although chemical changes that occur with age may be analyzed in the laboratory. There are no definite differences between the sexes, but evidence of dyeing, bleaching, lacquering or waving can possibly be a sign that the hair is female. Certain broad distinctions also exist between the hair from different races.

or outer sheath, formed of overlapping scales; the cortex, containing the pigment granules that give hair its natural colour; and the hollow medulla, the central core, which contains air.

The cuticle provides the easiest means of distinguishing human hair from that of other animal species, generally by the shape of the epidermal scales. All forensic laboratories keep recognition charts that enable the species to be quickly identified. The colour and distribution of pigment granules in the cortex are important in distinguishing the hair of a particular individual. The medulla is described, according to its appearance, as continuous, interrupted or fragmented – or, in rare cases, there may in fact be no medulla at all.

The comparison microscope (see "The Speeding Bullet") is the means normally used to show up similarities and differences between two sample hairs. A hair can also be examined in cross-section: it is embedded in a block of paraffin wax that is then cut into thin slices. These can then be transferred to a microscope slide.

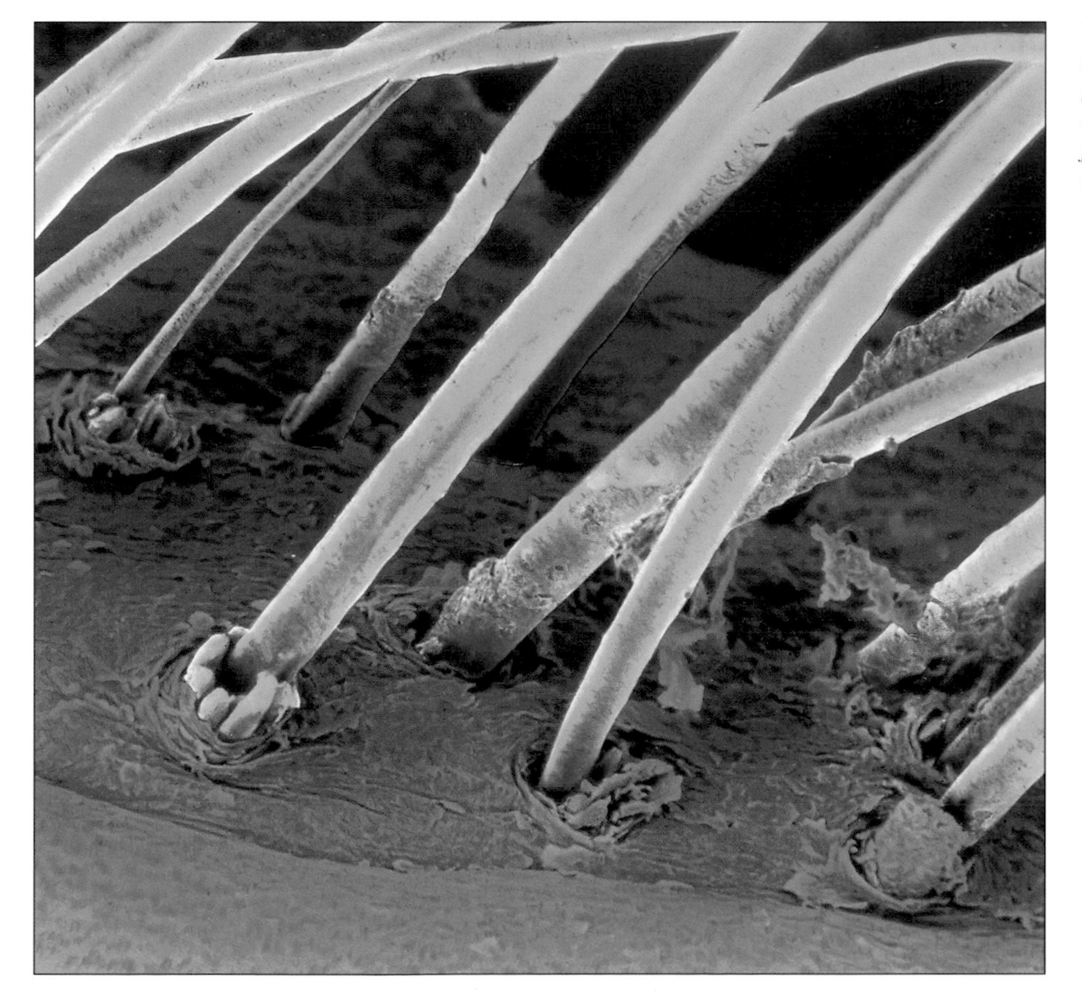

The human eyelash hairs are also generally circular in cross-section, but finer and smoother in appearance.

CRIME FILE:

Napoleon Bonaparte

For 140 years, rumours have persisted that the defeated French emperor, exiled to a tiny mid-Atlantic island, died of poison administered by agents of the English. Modern neutron activation analysis lends support to this belief.

The former French emperor Napoleon Bonaparte, who may have been poisoned with arsenic during his exile on the island of St. Helena.

When Napoleon Bonaparte died in exile on the island of St. Helena in 1821, his valet preserved a lock of his hair as a memento. Two months before, Napoleon had written: "I am dying before my time, murdered by the English oligarchy and its hired assassins", and this gave rise to persistent rumours that he had been poisoned.

With the development of neutron activation analysis, it was decided to submit some of the hair to test. Even after 140 years, the analysis revealed more than 10 parts per million of arsenic in the hair, considerably more than normal (see "With Poison Deadly"). In particular, the distribution of the arsenic along the shaft of the hair suggested that the deposed emperor had received a series of heavy doses over the four months preceding his death. Whether Napoleon had been poisoned by his English jailers, by a member of his entourage, or by medicines that he himself took during his imprisonment, is now impossible to establish.

CASE CLOSED

Comparing hair samples under the microscope, however, is a far from foolproof method of identification – indeed, the best that can be said is that two samples closely resemble one another. In cases of poisoning with antimony, arsenic, or thallium, these elements will appear in the keratin, and it is even possible, by analyzing successive short sections of hair, to estimate the times at which they are absorbed. Thallium also produces a characteristic shrinking of the hair root, which causes it to fall out. (See "With Poison Deadly".)

In the 1950s, Dr. Robert J. Jervis at the University of Toronto developed a system of neutron activation analysis. Each element present in the hair emits gamma-rays of characteristic wavelength when bombarded with neutrons. It has been claimed that the chances of sample hairs from two sources having exactly the same chemical makeup are about one in a million.

Hair that is pulled from the body, rather than falling out naturally, or being cut or broken, usually has particles of tissue from the follicle adhering to the root. This tissue can be analyzed for blood types or characteristic DNA. It is probable that hair samples will in future prove to be the prime evidence in many criminal trials, rather than corroborating other incriminating evidence.

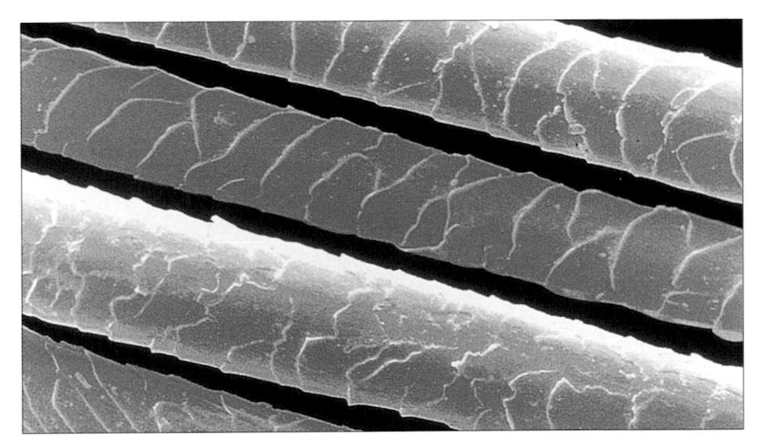

Dog hairs. The overlapping scales of the cuticle are noticeably different from those on human hair (see page 163).

Foreign Fibres

A knitted sweater, or any similar garment, sheds fibres constantly, and it also picks up "foreign" fibres by contact. Even a smoothly woven fabric, if for instance it rubs against a door, leaves behind a few traces. If a car hits a

The waves in a beam of light vibrate at all angles through 360°, but if the beam passes through certain materials it is "polarized", so that the waves vibrate in only one plane. This also happens when light is reflected: polarizing sunglasses and photographic filters are made from materials that let light through in only one plane, set at a different angle from that of the reflected light, so that dazzle is eliminated.

When a man-made material is forced through a spinneret, its molecules are aligned parallel to the length of the fibre. If a beam of polarized light is shone through the fibre, the alignment of these molecules causes it to emerge in two components: one in which the light vibrates parallel to the length of the fibre, and one in which it vibrates at right angles. These two travel at different speeds through the fibre, and emerge "out of phase".

Since the refractive index (see Chapter 14) measures the speed at which light passes through a transparent substance, the fibre has, therefore, two refractive indices: the difference between the two is known as the birefringence value. Measurement of this value enables the nature of the fibre to be determined.

In addition, infrared spectrometry can measure the absorption of different wavelengths of light passing through the fibre, providing a "signature" that can be compared with a reference set of "signatures" for all known fibres. Spectrometry can also be used to analyze the different dyes used in the manufacture of fabrics. To do this, a specialized miniature instrument is attached to a conventional microscope. Only minute quantities of fibre fragment are required. If there is still some doubt about the exact composition of an unidentified fibre, it can be analyzed by gas chromatography. It is heated to a high temperature until it decomposes into gaseous components that can then be separated and identified.

CRIME FILE:

John Francis Duffy

He was known as the "railway rapist" because his attacks often took place close to train stations. When he turned to murder, an unusual brand of string was to help secure his conviction.

In 1988, an English court found the "railway rapist" John Francis Duffy (see "The Guilty Party") guilty of the murder of two young women – and the rape of more than thirty others. Two items of fibre analysis played an important part in securing his conviction.

Duffy's first murder victim was 19-year old Alison Day, raped and killed on December 29, 1985. He threw her body into the river Lea at Hackney Wick, east London. It was found 17 days later, and her sheepskin jacket was also recovered from the river bed by a police diver. After it had been carefully dried out, fibres that might have come from the attacker's clothing were discovered on it, as well as other foreign fibres on Day's shirt and jeans.

CASE CLOSED

Just four months later, 15-year-old Maartje Tamboezer's body was found. Her hands had been tied with an unusual brown string, which was made from twisted paper. It was a brand known as Somyarn, and was manufactured at a factory in Lancashire; the manufacturers were able to explain that it had been made from an unusually wide-edged strip of paper, and had not been manufactured since 1982.

Duffy was eventually arrested in the autumn of 1986. When his mother's home was searched, a ball of Somyarn string was found beneath the stairs. When 30 items of his clothing were examined at the Metropolitan Police laboratory, and 2000 separate fibre samples taken, 13 fibres from Alison Day's clothing were found to match those from Duffy's clothes. "It was virtually a fingerprint," said Detective Sergeant Charles Farquhar afterwards.

Police searching the river bank where Alison Day's body was found in January 1986. She had been raped and killed by John Francis Duffy, the so-called "railway rapist".

CRIME FILE:

Andreas Schlicher

He murdered a young woman in Germany, cut off her head and concealed it. But his shoes bore traces of soil from the scene of the crime, and coloured fibres matched those from the victim's clothing.

Margarethe Filbert was reported missing on May 29, 1908: she had gone for an afternoon walk in the Falkenstein valley in southern Bavaria in Germany, and failed to return. The following day her headless body was found in the woods. At first the police thought that she was the victim of a sexual assault – she lay on her back with her skirt and petticoat pulled up over her body – but the autopsy revealed that there had been no rape, and leaves trapped in the clothing suggested that she had been dragged by the legs through the undergrowth. The pathologist concluded that she had been strangled, and then beheaded with a knife.

Hairs were found clutched in Filbert's hand. The local authorities consulted Georg Popp, a Frankfurt analytical chemist who had been involved in several criminal investigations. He reported that the hairs came from a woman's head. Without Filbert's head it was impossible, at a date long before DNA typing, to establish whether the hair was hers.

Suspicion fell on a local farmer, Andreas Schlicher, who had a reputation for violent behaviour. Traces of human blood were found on his clothing and under his fingernails, and there was other circumstantial evidence, but not sufficient to establish his guilt.

Popp was then given Schlicher's shoes to examine. He found various layers of soil that he identified with the route that Schlicher would have taken from the scene of the crime. He also discovered fragments of wool and cotton fibre, some purple and some reddish-brown, that matched the materials of the victim's skirt and petticoat.

Employing a spectrophotometer, Popp identified the coloured dyes in the fibre fragments, and established that they were identical with those used in the clothing. This evidence proved sufficient for a jury to find Schlicher guilty. After the trial he finally admitted to committing the crime and described where he had hidden Margarethe Filbert's head.

CASE CLOSED

pedestrian, the bodywork is almost certain to retain fibres from clothing, which can be discovered with a hand lens and removed with sticky tape for examination.

The first tool available to the forensic scientist in the examination and identification of fibres is the comparison microscope. Forensic laboratories maintain comprehensive catalogues of natural and artificial fibres, and provisional identification can be made relatively easily.

Man-made fibres have a very different structure from natural ones. They are produced by forcing a liquid under pressure through the fine holes of a machine head called a spinneret, and as a result they have an external

CRIME FILE: Wayne Williams

More than 20 young males in Atlanta fell victim to his murderous homosexuality before he was caught. Eventually, the FBI identified carpet and other fibres from his parents' home, and the hairs of his German shepherd dog.

Wayne Williams, who, over a 2-year period,murdered more than 20 young African-Americans on the outskirts of Atlanta, Georgia.

Between July 1979 and May 1981, more than 20 African-American males were found asphyxiated or otherwise murdered on the outskirts of Atlanta, Georgia. Investigators discovered identical fibres on their clothing, suggesting that many of these were the work of a serial killer. The fibres were examined at Georgia State Crime Laboratory, and they turned out to be of two types: a yellowish-green nylon that appeared to be from a carpet, and a violet-coloured acetate. When news of this was published in a local paper in February 1981, the killer immediately changed his habits, beginning to dump his victims, stripped of most of their clothing, in rivers. One of these was Jimmy Lee Payne, found dead on April 27: on his shorts was a single rayon fibre.

Police decided to stake out a bridge over the Chattahoochee River, and during the night of May 22 they heard a loud splash. A second patrol car, alerted, stopped a station wagon driven by Wayne Williams, a 23-year-old African-American music promoter. He was questioned – he said he had just dumped some garbage into the river – and then allowed to leave.

The body of Nathaniel Cater was dragged from the Chattahoochee two days later: a yellowish-green carpet fibre was found in his hair. On June 2 the police obtained a warrant to search Williams's family home. It was carpeted throughout with yellowish-green nylon, which matched the fibres obtained from the bodies.

In itself, this evidence was insufficient to connect Williams with the killings, and the examination of the fibres was turned over to the FBI laboratory. Birefringence analysis showed that the nylon was manufactured by the Boston company of Wellman Inc. The particular type had been made between 1967 and 1974, and sold to various carpet makers. Identification of the dyes led to the West Point Pepperell Corporation of Dalton, Georgia. The carpeting was known as Luxaire, and its colour as English Olive. The Wellman fibre had been used only between 1970 and 1971, and the carpet sold throughout 10 south-eastern states, including Georgia.

What were the odds that the fibres found on the bodies came from the carpet in the Williams household? Assuming an even distribution in the 10 states, knowing the total area sold by West Point Pepperell, and the total number of homes – nearly 640,000 – in Atlanta City alone, the FBI calculated the chances at 1 in 7792.

When Williams's car was searched, fibres from the carpet matched the rayon found on Jimmy Payne's shorts. A further statistical calculation put the likelihood of this being by chance at 1 in 3828. Combining these figures put the odds at nearly 24 million to 1. Also found in the car were violet acetate fibres that matched both a blanket in Wayne's bedroom and the fibres from the earlier victims' clothing.

Despite the difficulty in explaining the statistical calculations to a jury – the prosecution prepared 40 charts and 350 photographs – Williams, charged only with the murders of Payne and Cater, was found guilty on February 27, 1982, and sentenced to two terms of life imprisonment.

The Chattahoochee River bridge, where Williams was questioned after he had disposed of one of his victims' bodies.

CRIME FILE:

Jeffrey MacDonald

He was found guilty of the murder of his wife and two daughters, but a retrial was sought. The appeal centred on blonde wig hairs found at the scene, but they were shown to come from Barbie dolls.

Captain Jeffrey MacDonald, the army doctor found guilty of stabbing his wife and two young daughters to death at Fort Bragg, North Carolina, in February 1970.

On the night of February 17, 1970, the military police at Fort Bragg, North Carolina, responded to an emergency call and found an appallingly bloody scene at the home of army doctor Captain Jeffrey MacDonald. His wife Colette was dead, stabbed 21 times; MacDonald himself, covered with bleeding wounds, was conscious but motionless; and daubed in blood on the bed-head was the word PIG. In an adjoining bedroom, their two young daughters had been stabbed and beaten to death.

MacDonald said he fell asleep on the living-room couch to be awoken by his wife's cries. He found four hippies standing over him, led by a woman wearing dark clothes, a floppy black hat, and a long blonde wig, who chanted "acid is groovy … kill the pigs". They slashed at him with a knife and an ice pick, until he fell unconscious. When he came to, he discovered the carnage in the bedrooms.

Army investigators found a wealth of evidence that cast doubt on MacDonald's tale, and on May 1 he was charged with murder. However, the investigation had been badly bungled, and many important items of trace evidence had been lost. In October all charges were dropped.

MacDonald resigned from the army, and his subsequent behaviour aroused the FBI's suspicions . They re-examined all the available evidence, which was presented to a grand jury in July 1974. MacDonald was charged with all three murders, was finally found guilty on July 16, 1974, and sentenced to three terms of life imprisonment.

MacDonald persistently appealed against the sentence, and in 1992 the distinguished lawyer Alan Dershowitz applied for a retrial. The basis of the appeal was that blonde wig hairs found on his wife's hairbrush had not previously been produced as evidence, and that they gave credence to his claim that a woman in a blonde wig had been in his home.

Again the FBI examined the trace evidence. They found two types of wig hair, one of which they had not seen before. Spectrometry revealed it to be saran, a fibre used for doll hair and dust mops. Eventually, the FBI found two examples of Barbie dolls with hair made of the fibre. They could not show that the two murdered little girls had possessed similar dolls, but they could say with confidence that the fibres did not come from a human wig.

Dershowitz raised the question of other unexplained fibres and hairs that had been discovered. The FBI showed that some came from a hairpiece possessed by Colette MacDonald, others from her clothing; and a hair on her body was MacDonald's. The appeal was disallowed.

CASE CLOSED

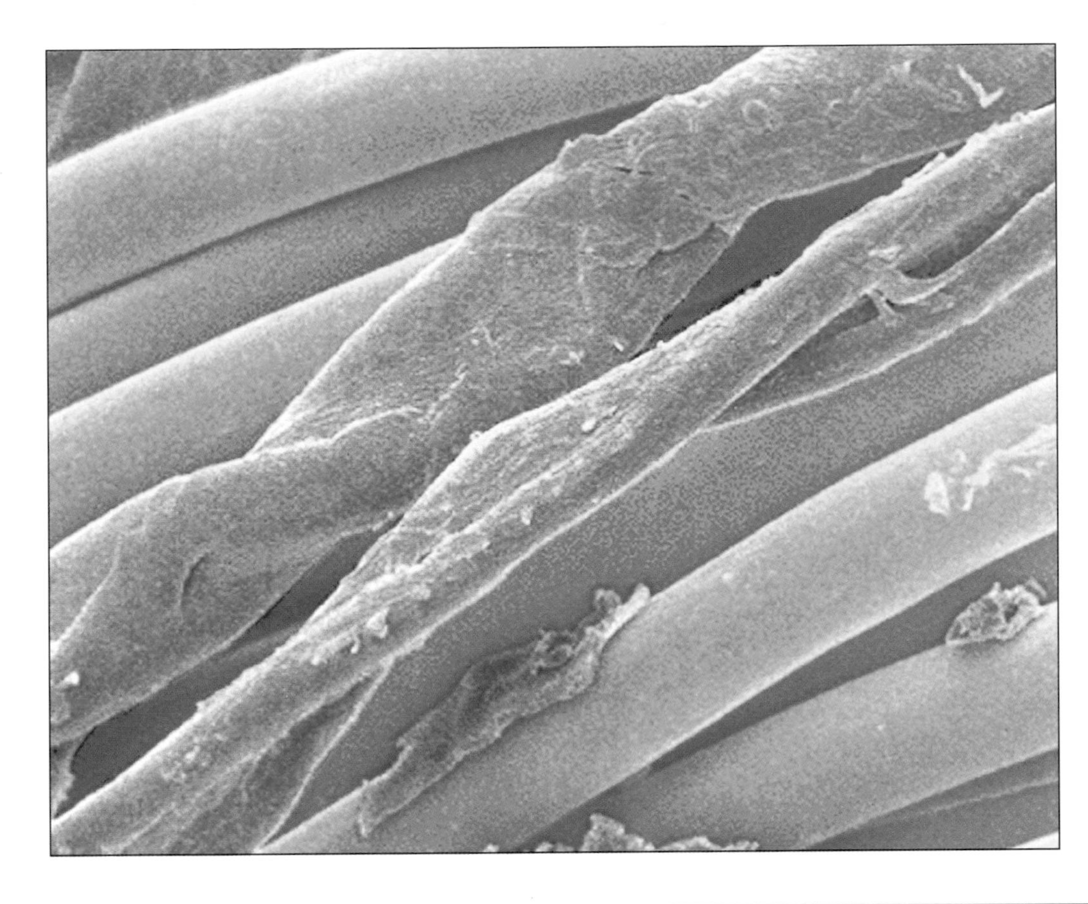

Left: A micrograph comparing natural cotton fibres (green) with synthetic polyester fibres (yellow). The synthetic fibres are smooth, without any structure.

Below: Extruded fibres of artificial silk, illuminated by polarized light. The light reveals differences within the physical structure of the fibres, but these are easily distinguishable from the structure of any natural fibre.

smoothness and regularity quite unlike that of natural fibres, and little or no internal structure when viewed in cross-section.

The particular material of which a man-made fibre is made can sometimes be determined by its solubility in various solvents, but the standard method for identification exploits a phenomenon known as "birefringence" (see page 165).

Equipped with a comparison microscope, a micro-spectrophotometer and a catalogue of fibre characteristics, the forensic scientist can measure the diameter of a fibre, determine the shape in cross-section, the birefringence value and spectrometric signature, observe the external characteristics along its length, analyze its colour, and identify it with a very high degree of certainty.

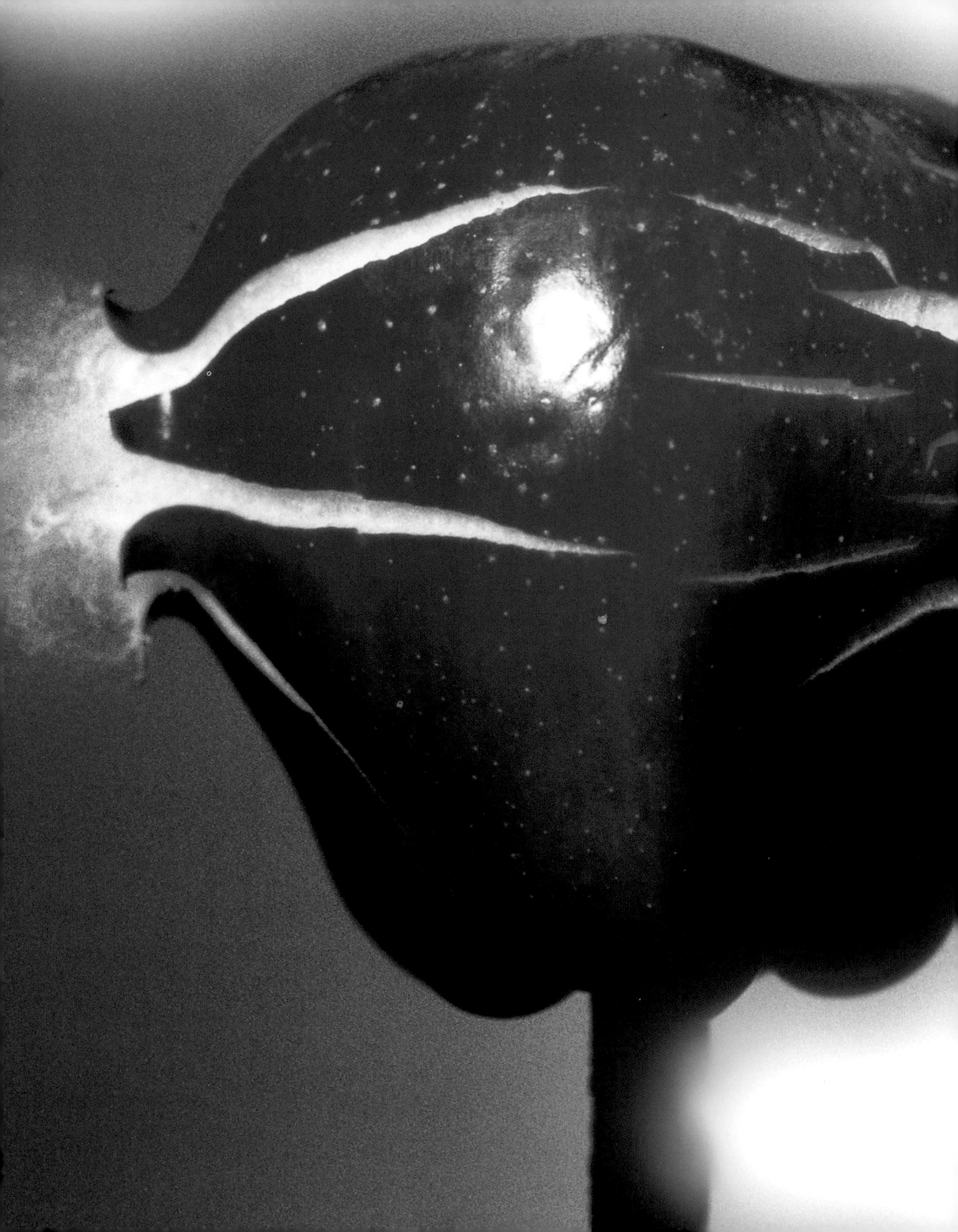

The Speeding Bullet

IT IS SOMETIMES BELIEVED THAT FORENSIC BALLISTICS – the examination and identification of guns and bullets – is a 20th-century science. This is true of the techniques that are routinely used today by crime laboratories in cases of gunshot injuries and deaths. History, however, has recorded some earlier successes.

In late 18th-century England, a Lancashire man named Edward Culshaw was shot dead. A certain John Toms, who owned a muzzle-loading pistol, was suspected of his murder. To use a pistol of this type, one first loaded it with gunpowder, then rammed a wad of paper on top. Next, the ball was inserted, and a second wad of paper was rammed on top to prevent it falling out. The surgeon who probed Culshaw's wound removed not only the ball, but a scrap of paper. It had been torn from the edge of a ballad sheet, the rest of which was found in Toms's pocket. The torn pieces matched exactly, and this evidence was sufficient to send Toms to the gallows.

A high-speed photograph of a .22 bullet fired through an apple demonstrates the terrible damage that it can do to human tissue. The exit wound is larger, bursting the skin outward in a star shape, and ejecting the pulp.

An investigation rather closer to modern ballistics was carried out by Henry Goddard of the Bow Street Runners (the precursors of England's first police force) in 1835. At Southampton, Hampshire, an apparent break-in and burglary occurred, during which a butler claimed to have been shot at as he lay in bed. Goddard examined the bullets from the butler's pistol, and compared them with the one dug out of the man's headboard. All showed the same characteristic: a tiny raised bump, caused by an imperfection in the mould from which they had been cast. It was clear that "the butler did it"; and he confessed to having set up a fake burglary.

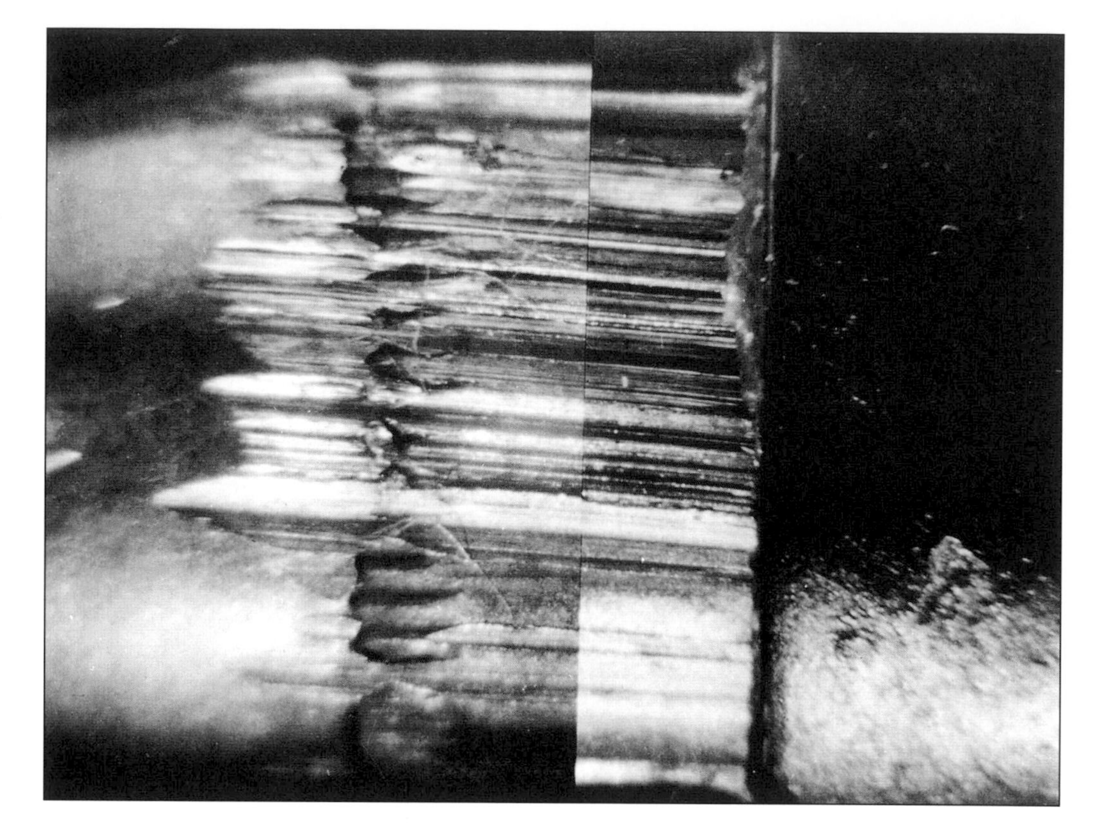

The spiral "lands" and grooves of a rifled gun barrel leave characteristic striations on the body of the bullet as it spins. No two guns, even from a single manufacturer, produce identical striations, and ballistics examiners can prove the identity of striations on a bullet recovered at the scene of a crime, and a test bullet fired from a suspect's gun.

French scientists took a different approach in 1869, when they analyzed the melting-point, weight and composition of a bullet taken from a murder victim's head. In this way they proved that it was identical with bullets found in possession of the suspect.

IDENTIFYING THE GUN

With the exception of smooth-bore shotguns, most guns since the early 19th century have been rifled. Spiral grooves cut into the inside of the barrel cause the bullet to spin and give the gun greater accuracy. The uncut parts of the barrel between the grooves are called "lands". Because bullets are made slightly larger than the bore of the gun, to ensure a tight fit, these lands produce clearly visible grooves down the length of the bullet.

This fact was first made use of in a murder case by Alexandre Lacassagne, professor of Forensic Medicine in the University of Lyon, France. In 1889 he matched seven grooves found on a bullet recovered from a victim's body with the seven rifling grooves in a gun in the possession of a suspect. The modern science of ballistics had arrived.

Types of rifling differ markedly from manufacturer to manufacturer. A firearms examiner can very quickly determine the make of gun from the number of rifling grooves, the relative widths of the lands and grooves, and whether the rifling spins the bullet in a right-handed or left-handed way.

As gun after gun is rifled on the same machine, the tools that cut the grooves become very slightly worn, or even damaged. This results in tiny imperfections in the bottom of the groove, and in each gun these will differ slightly. These imperfections produce slight scratches (striations) on the bullet, parallel to the grooves cut by the lands, that are characteristic of the gun from which it has been fired. Minute examination of the striations under a microscope can therefore make it possible to identify an individual gun.

Sometimes, if automatic or semiautomatic weapons are used, spent cartridge cases are found at the scene of a firearms crime – although careful criminals will always try to gather them up and carry them away. These, too, can provide evidence of the particular gun that fired them.

The bullet lies against the breech block of the gun, which is made of hardened steel. Pulling the trigger drives the firing pin through a small aperture in the breech block to strike the detonator cap of the cartridge. As the gun fires, the immense pressure generated drives the case against the breech block. As a result, the softer metal of the cartridge becomes imprinted with any imperfection – produced during manufacture or subsequent use – in the steel. At the same time, the firing pin leaves the shape of its own print on the cap. The ejector mechanism is also likely to leave characteristic marks on the cartridge case.

Judge Oliver Wendell Holmes. In 1902, he had a gunsmith fire a test bullet for comparison with one recovered at the scene of a murder. As Holmes stated in his judgment: "I see no other way in which the jury could have learned so intelligently how a gun barrel would have marked a lead bullet fired through it."

Given a lone bullet – provided it is not too distorted by impact with its target – the ballistics expert can confidently describe the make, and the details of the barrel through which it was fired. Given the cartridge case, he or she can predict the characteristics of the breech block, firing pin and ejector. When the suspected weapon is found a case can be built up.

It is possible to examine the interior of a rifled barrel – in the mid-1920s the American John H. Fischer invented the "helixometer", based on the medical cytoscope, for just that purpose. It is more usual, however, to compare bullets from the crime scene with bullets fired from the suspect gun under test conditions.

In 1900, Dr. A. Llewellyn Hall published *The Missile and the Weapon*, a book devoted to the problems of firearm identification, which came to the attention of the great American jurist Oliver Wendell Holmes. In 1902, Holmes heard a trial in which a man named Best was accused of murder with a

CRIME FILE:

Nicola Sacco and Bartolomeo Vanzetti

The trial and conviction of the two "anarchists" for murder became a cause célèbre in 1920s America, and across the world. At retrial, ballistics experts demonstrated that the fatal bullet had definitely come from Sacco's gun.

On the afternoon of April 15, 1920, two men leapt from a Buick outside a shoe factory at South Braintree, Massachusetts, shot down two security guards, and made off with nearly $16,000 intended for the company's payroll. Eyewitnesses described the men as "Italian-looking", and reported that there seemed to be as many as three other men in the car. Around the guards' bodies lay a number of .32 bullet shells, which had been manufactured by three firms: Peters, Winchester, and Remington.

A little later, two men were arrested on a streetcar in nearby Bridgewater. They were immigrant Italians: 29-year-old Nicola Sacco and 32-year-old Bartolomeo Vanzetti. Sacco was carrying a loaded .32 Colt revolver, together with 23 bullets, variously manufactured by Peters, Winchester, and Remington. In addition, Vanzetti was armed with a revolver, a .32 Harrington-Richardson, and carried four shotgun cartridges identical with one found at the scene of a failed payroll snatch four months earlier.

For some years, America had suffered a number of "anarchist" outrages, often involving home-made bombs. The fundamental cause of the violence was industrial unrest in the east coast states,

(Above) Sacco (right) and Vanzetti (left) arriving at court for their trial. (Below) One of the many demonstrations provoked by the arrest of the two "anarchists".

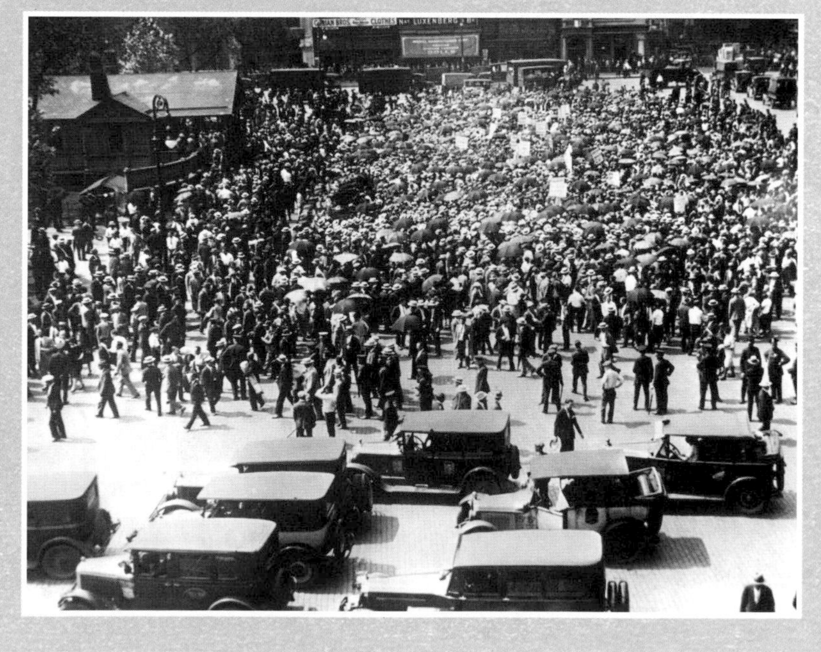

due to the poor conditions of employment among immigrant workers. The arrest of the two Italians showed that something positive was being done by the police – and it also gave the workers a justification for protest.

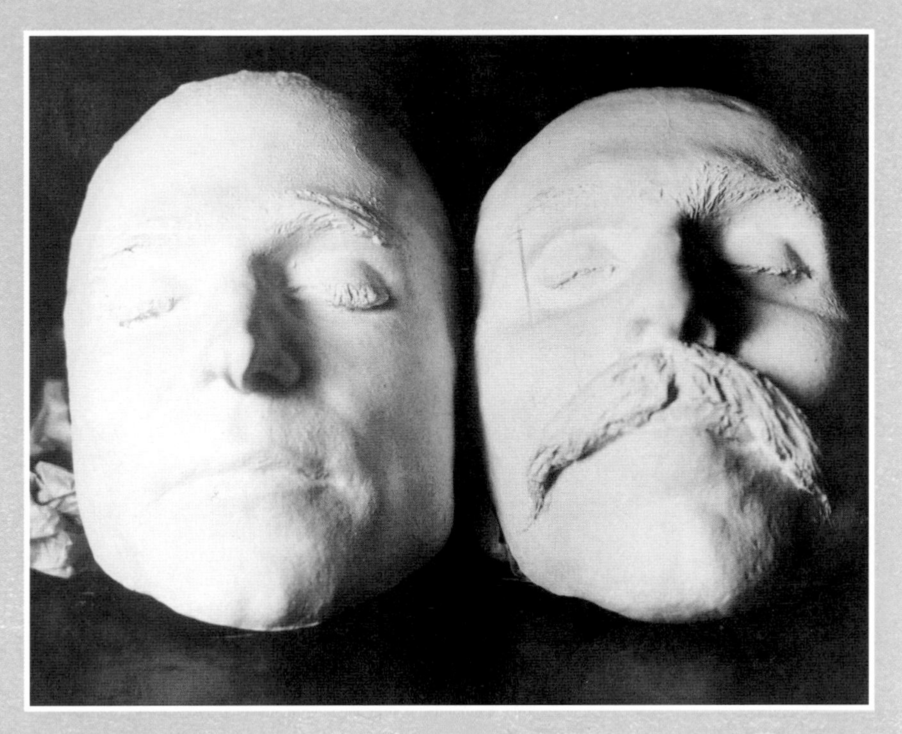

Vanzetti was found guilty of the first payroll attempt and sentenced to 15 years in prison, but Sacco had a firm alibi. The trial of both for the double murder at South Braintree opened on May 31, 1921. Meanwhile, their defence team had sought support from left-wing organizations all over the world, and established a fund under the name of the Sacco-Vanzetti Defense Committee. It began to seem that the two men were being tried more for their politics than their crime.

Witnesses for the prosecution totalled 59, those for the defence 99. One prosecution expert testified that Sacco's gun had fired the fatal shots, but two defence experts, James Burns and Augustus Gill, declared that this was not possible. A damning piece of evidence was that the bullets were of obsolete manufacture: none could be found except the 23 in Sacco's pocket. In July, the jury returned a verdict of guilty, and both men were sentenced to death.

International protest continued, and the defence appealed for a retrial. Prosecution expert Charles Van Amburgh re-examined the ballistics evidence and, making use of the technological advances developed by Charles Waite, produced photographs of one fatal bullet and test bullets fired from Sacco's gun. Finally, in June 1927, Colonel Goddard offered his services as an unbiased expert. With Gill as a witness, he fired a test bullet into cotton wool, then placed it beside the other under his comparison microscope. There was no doubt, as Gill had to agree: "Well, what do you know about that!" he exclaimed. When Burns also changed his opinion, there was no further hope for Sacco and Vanzetti. They went to the electric chair on August 23, 1927. Vanzetti's last words were "I am innocent"; Sacco cried "Long live anarchy!"

After their execution on August 23, 1927, death masks were made of Nicola Sacco and Bartolomeo Vanzetti.

But the controversy lingered on. More than 30 years later, in October 1961, a forensic team led by Colonel Frank Jury, former head of the New Jersey Firearms Laboratory, concluded that the bullet had indeed come from Sacco's gun. Finally, in March 1983, another team, paid for by a Boston television station, confirmed Goddard's findings.

CASE CLOSED

The aftermath of the "St. Valentine's Day massacre" of 14 February 1929, when members of Al Capone's gang, led by "Machinegun Jack" McGurk, and disguised as police, slaughtered George "Bugs" Moran and six of his men.

revolver, and he called in a gunsmith to examine the evidence. The gunsmith fired a shot from Best's gun into a box of wadded cotton wool; in court, with a magnifying glass, he demonstrated the points of similarity between the two bullets to the jury.

The same method of testing is employed, in principle, today. Because the cotton fibres themselves may mark the bullet, it is now more usual to fire test shots into a water tank. Approximately 6 feet (1.8 metres) of water is sufficient to stop the bullet.

The Comparison Microscope

The science of forensic ballistics was firmly established in the United States by the work of Charles Waite, an assistant in the office of the New York State Prosecutor. In 1915, a German immigrant named Stielow was found guilty of the shooting of 70-year-old farmer Charles Phelps and his housekeeper, and sent to Sing Sing to await the electric chair. His attorneys obtained a stay of execution in July 1916, and shortly after two vagrants confessed to the crime. Waite asked a member of the New York City homicide squad, Captain Jones, to examine Stielow's gun. Jones reported that it was so badly corroded that it could not have been fired for four or five years. Test shots showed marked differences between bullets from this gun and those taken from the bodies of the victims; finally microscopic examination at a laboratory in Rochester established that the grooves left by the lands were not at all similar. Stielow was given a free pardon.

After serving in the army during World War I, Waite spent two years travelling throughout the United States and Europe, collecting all sorts of data from every firearms manufacturer. His work soon led to the establishment of the Bureau of Forensic Ballistics in New York, the first of its kind in the world. When Waite died in 1926, he was succeeded as head of the Bureau by Colonel Calvin Goddard. One of Goddard's first major cases was that of Sacco and Vanzetti.

The next development in ballistics was the invention of the comparison microscope by Waite's collaborator, the chemist Philip O. Gravelle. Essentially, this consists of two microscope objectives with a single eyepiece: when a bullet is placed below each objective, the marks on both can be minutely compared.

In 1929, with the aid of this instrument, Goddard identified the two Thompson sub-machineguns used in the notorious St. Valentine's Day massacre in Chicago. This success so impressed J. Edgar Hoover, director of the recently formed FBI, that he persuaded Goddard to found the Scientific Crime

CRIME FILE: Las Vegas Crash

When a plane crashed without apparent cause, the FBI were called in. An examiner discovered positive traces of lead – undoubtedly from a bullet – in the framework of the pilot's seat.

In the 1950s, an American plane carrying a party of gamblers from San Francisco to Las Vegas crashed without any apparent reason, and it was suggested that the pilot might have been shot. The FBI sent agent Bill Magee to the crash scene to investigate, but he found that the thousands of rivet holes in the pieces of the plane's fuselage almost exactly matched the hole that would have been made by a .38 bullet. There seemed no easy way of discovering whether a gun had been fired aboard the plane.

Later, while demonstrating the chemical test for lead traces to other agents, Magee picked up a piece of metal tubing with a dent in it, and found to his surprise that it gave a positive reaction. The tubing turned out to be part of the pilot's seat, proving – at the very least – that a shot had been fired at him.

CASE CLOSED

Detection Laboratory at Northwestern University, Evanston, Illinois. Not long after, Hoover set up the FBI Ballistics Department in Washington DC. This is now probably the largest and busiest firearms investigation centre in the world.

In Britain, news of the comparison microscope attracted the attention of Robert Churchill, a 41-year-old gun maker in London, who had been appearing as an expert witness in firearms cases since 1912. He and his collaborator, Major Hugh Pollard, had been experimenting with a similar instrument, based upon a rather primitive version developed as early as 1919 by the Scottish pathologist Sydney Smith. In 1927 Churchill travelled to the United States, where he met Colonel Goddard, and gathered sufficient information on the comparison microscope to have one made to his specifications when he returned to London.

In September 1928, Churchill had his first success with the instrument. Two men, Frederick Browne and William Kennedy, were on trial for the murder of a police constable, George Gutteridge. Churchill led a team of three experts from the War Office, who gave evidence that striations on the bullet that had killed Gutteridge exactly matched those on bullets fired under test from Browne's Webley revolver. They also testified that marks on the cap of a cartridge case found after the crime matched those on the breech block. They reported that they had test-fired no less than 1300 revolvers of the same type without discovering a similar mark. Browne and Kennedy were found guilty, and sentenced to hang.

The case, coming so soon after the final verdict on Sacco and Vanzetti, attracted international attention. Within a few years, special ballistics laboratories were established at Lyon, in France; at Stuttgart and Berlin, in Germany; at Oslo, in Norway; and not long afterward in Moscow. The science of forensic ballistics had come of age.

Patterns of Shot and Powder

Some months before the Gutteridge case, Churchill had provided a very different sort of evidence in another British gunshot murder. On the night of October 10, 1927, a poacher named Enoch Dix took his .410 single-barrelled shotgun into Whistling Copse, on the estate of Lord Temple, near Bath, Somerset. He was spotted by the head gamekeeper, William Walker, and his under-keeper George Rawlings, who pursued him. Dix spun round, and his gun fired: Walker fell dying, and Rawlings took a quick shot at the fleeing poacher.

When police searched Dix's cottage they found the gun, and discovered that his back was peppered with pellet wounds. He claimed that Rawlings had fired first, and that his gun had gone off accidentally with the shock.

Churchill was asked for his advice on who had fired first, and at what range. He took both guns, and identical cartridges charged with the same size shot, and fired at a succession of whitewashed steel plates. At 15 yards (13.6 metres), he found, the spread of shot was between 27 and 30 inches (68.6 and 76.2 centimetres); at 20 yards (18.2 metres), it was between 36 and 38 inches (91.4 and 96.5 centimetres). From the wounds on Dix's back, and pellets that had hit a nearby tree, Churchill calculated that Dix must have been at least 15 yards (13.6 metres) away when Rawlings fired. If his gun had then gone off, Walker would have been similarly peppered – but his fatal wound covered only some 4–5 inches (10.2–12.7 centimetres), and he had evidently been shot at almost point-blank range. Despite the judge's directions to the jury, Dix was found guilty only of manslaughter.

A human leg showing the typical patterning of wounds from the pellets of a shotgun.

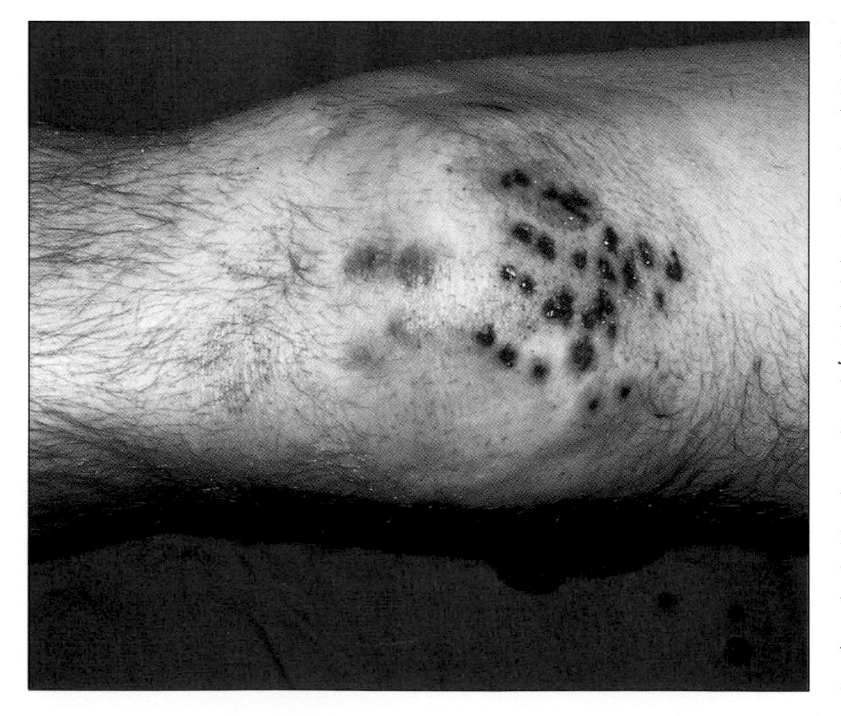

Since then, the close examination of gunshot wounds has proved important in many cases. When a weapon is fired at close range, burned – and sometimes unburned – particles of gunpowder produce a characteristic "tattooing" on bare skin around the wound, and a similar pattern on clothing. The size of this deposit can be a guide to the distance and direction from which the shot was fired. If the gun was aimed directly at the victim, the pattern is almost circular. If the weapon is held in contact with the skin, or up to half an inch (a centimetre) away, the powder pattern is usually absent. The same is true of shots from a distance greater than 10 feet (3 metres).

When a bullet passes through any kind of material, it leaves minute traces of lead around the hole, which can be detected with chemical reagents. A bullet is also extremely hot, and a close-range shot can melt synthetic fibres in the clothing. These observations, too, can give an indication of distance. Quite often the suspect claims that the gun went off accidentally during a struggle. If this is true, evidence of the pattern of residues may substantiate or disprove the claim.

A bullet picks up traces of anything it passes through, or ricochets from: fragments of bone, hair, wall materials, paint, glass, fibres – even blood. These can be valuable in calculating the bullet's path. When a Pennsylvania police officer was accused of killing an innocent motorist, he claimed that he had tripped, and fired his gun by accident. Examination of the bullet under the microscope showed traces of both cement and glass, proving that it had ricocheted from the roadway, and gone through the car window to strike the victim.

The very rapid expansion of high-pressure gases in a gun barrel is followed by a "blowback", which can cause material to be drawn from the surroundings into the barrel. In a murder case in Florida, the killer held a pillow over the gun's muzzle, in an attempt to muffle the sound of the shots. A suspect was subsequently arrested, and fragments of a feather were found in his gun.

When a suspect is detained at the scene of a shooting, or shortly afterward, it is standard practice to examine his or her hands. Tiny residues from the primer explosive in the detonator will be blown out of the breech, or carried on to the hands by blowback, and swabbed samples can reveal whether the suspect has recently fired a gun. Formerly, the swab was tested for nitrates. Since nitrates are increasingly used in cosmetics and cigarettes, as well as agricultural chemicals, other tests are now employed. Barium and antimony from the primer, for example, form microscopic particles, which can be observed by electron microscope, and identified by chemical reagents.

Where Did the Bullet Go?

A spinning bullet leaves a rifled barrel at over 1500 feet (450 metres) per second, and as it does so it develops "tail wag" – not unlike the wobble of a child's spinning top before it settles down to steady rotation. This tail wag can create an entrance hole bigger than the calibre of the bullet itself. Inside the target, the bullet can be deflected in such a way that – even if it is not damaged – the exit

CRIME FILE:

Lee Harvey Oswald

How many shots were fired in Dealey Plaza when President Kennedy was assassinated in 1963? Forensic pathologists, examining the evidence in 1977,were convinced there had been only two.

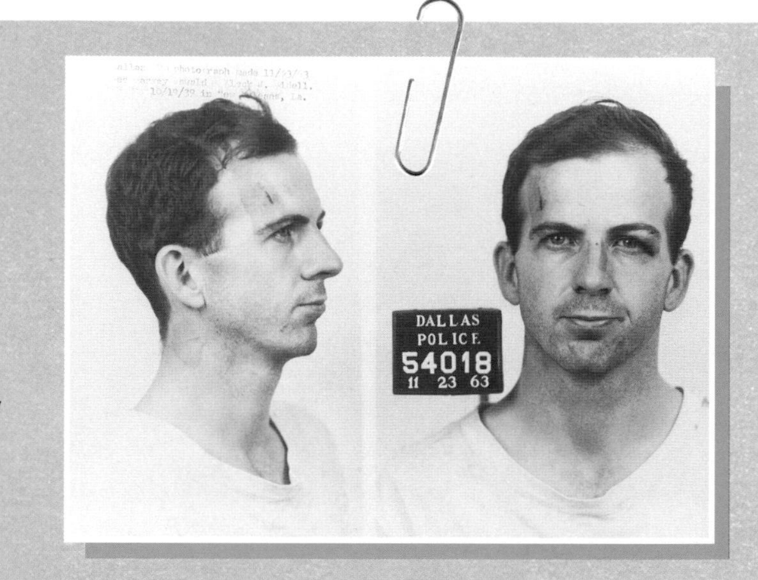

Lee Harvey Oswald, photographed at Dallas police station following the assassination of President John F. Kennedy on November 22, 1963.

The near-panic that was to followed the assassination of President Kennedy in Dallas, Texas, on November 22, 1963; the forcible removal of his body, by the FBI, from Parkland Hospital in Dallas to Bethesda Naval Hospital in Washington DC; the secrecy surrounding some of the autopsy X-rays and specimens, and the disappearance of others – all these helped to fuel a wealth of conspiracy theories.

None of the medical pathologists who examined the president's body, both in Dallas and in Washington, had any experience of gunshot wounds. `The Warren Commission, set up in 1964 to examine the evidence and dispel the rumours, did not interview a single forensic pathologist. It was not until 1977 that the Congress Select Committee on Assassinations assembled a panel of forensic pathologists to review the evidence. Under Dr. Michael Baden, the New York City Medical Examiner, the panel looked again at the medical and autopsy reports, photographs, X-rays and the president's clothing.

One of the first questions the panel had to resolve was the number of shots fired at the presidential motorcade, and the direction from which the shots came. When Commander James Humes, the Navy pathologist, had examined the president's body at Bethesda in 1963, he had found a wound in the back, a massive head wound and what appeared to be a large entry wound in the front of the throat. X-rays showed no bullets inside the body. It seemed that the bullet that had struck the back had gone in a few inches and afterwards fallen out of the same hole

The sniper's hiding place on the sixth floor of the Texas School Depository Building, Dallas.

The sequence of events: (1) President Kennedy speaks to his wife (2) The president is hit (3) Mrs. Kennedy puts her arm around him (4) Mrs. Kennedy climbs onto the car trunk to seek aid and Secret Service man Clinton J. Hill jumps onto the car to help (5) Hill pushes Mrs. Kennedy back into the car (6) He then shields her and the president with his body, and the car speeds off to the hospital.

by which it had entered. But, as Humes told the FBI, this was impossible – the cavitation effect would have prevented it. Neither could Humes understand what had happened to the bullet in the head. No examination of the tissues was made to determine the tracks of the two bullets.

The next day, when JFK's body had been taken away for burial, Humes had telephoned Dr. Malcolm Perry in Dallas, and learnt that a tracheotomy had been performed on the president in an attempt to allow him to breathe. This had obscured the exit wound of the bullet that had entered the back. Because the bullet was a metal-jacketed military round, it had then struck Governor John Connally in the back sideways on, just above the right armpit, injured his lung and fifth rib, exited below his right nipple, then entered his right wrist through the radius bone, and finally passed through part of his left thigh.

Examination of the president's clothing confirmed this: the shirt and jacket each had a neat round hole in the back, and there were slit-like exit holes in the tie and shirt collar. Finally, when Connally allowed Dr. Baden to examine his back, his wound scar was revealed as being 2 inches (5 centimetres) long, clear evidence that the bullet had been moving sideways. The bullet that had done all the damage was found lying on the stretcher that carried Connally to hospital, where it had fallen out of the wound in his thigh.

As for the bullet that had struck the president in the head, the forensic experts made enhanced prints of the available X-rays, and showed the track of the bullet, which had entered an inch or two (a few centimetres) below the crown, and made a massive exit wound above his right ear. That bullet had struck the windscreen pillar of the car, and was found on the floor. Dr. Baden and his colleagues were convinced that there had only been two shots, and that both had come from behind.

CASE CLOSED

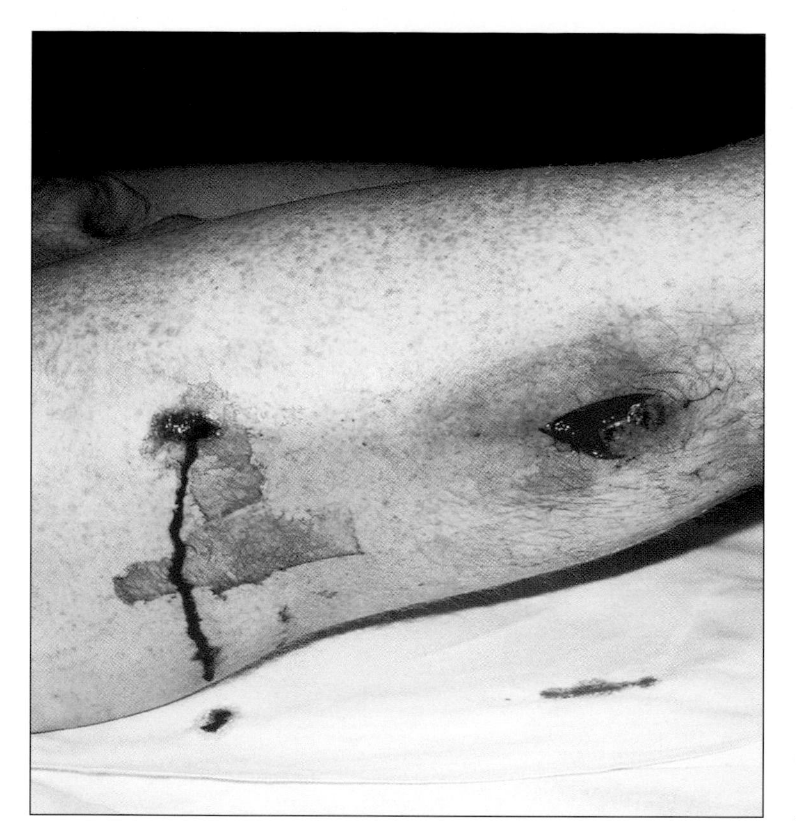

The wound caused by a bullet's exit from the body (right) is usually larger than the entry wound.

hole can be many times bigger than the entrance.

Frequently, the entrance wound is a small, clean hole, with an "abrasion collar", due to the frictional heat of the bullet where it penetrates the skin. Provided the gun has not been discharged at point-blank range – in which case the hole may be smaller than the bullet that caused it – the size of the hole can provide an approximate measure of the calibre of the bullet. At longer range, the bullet may already be tumbling, and cause a large lacerated wound.

Within the tissues, "cavitation" occurs: the force of the bullet causes the tissues to expand and then collapse in upon themselves, leaving a clearly detectable track.

The exit wound is usually larger, bursting the skin outward in a star shape, and if the bullet has struck bone – or even other tissues – it may disintegrate, partially or wholly, tearing a large hole. If, however, the skin is restricted by a belt or other tight clothing, or even if the victim is against a wall, the exit wound may be as small as the entrance wound.

Given all these factors, it is clearly not easy to determine with any certainty the exact direction from which the bullet was fired, nor its calibre. Analysis of any fragments may help to indicate the kind of bullet, and possibly its approximate size, but the behaviour of any bullet, both during its flight and after it penetrates its target, is difficult to determine. Very careful examination of the victim can sometimes provide the essential clue.

Once inside the victim's body, a bullet can travel in many strange ways. Two cases described by the Scottish pathologist Sir Sydney Smith exemplify this. In one, an Army deserter was shot while he resisted arrest, and died from haemorrhage shortly after. The bullet entered the outer side of his left thigh, leaving a clean entrance wound. It then passed through the flesh behind the femur, pulping the muscle, but leaving the major blood vessels undamaged, and emerged, creating an exit wound nearly 3 inches (7.5 centimetres) in diameter. The bullet then entered the inside of the right thigh, causing a lacerated wound 3 by 6 inches (7.5 by 15.2 centimetres) in size. After destroying more muscle tissue, it struck the lower end of the femur and disintegrated, powdering the bone and severing the main artery. A fragment or two left a small exit hole in the outside of the thigh.

"Anyone without experience or knowledge of the circumstances of the shooting," wrote Smith, "might, on looking at the wounds, have assumed that two shots had been fired, one from the left and the other from the right."

In the second case, a young soldier was badly wounded in both arms and both legs, each revealing an entry and an exit wound. It transpired that he had been bending forward to adjust his gaiters when his neighbour's gun accidentally fired. The bullet entered the outer side of his left leg below the knee, passed through his left arm below the elbow and into his right leg, finally entering his right arm. Very little damage was done by the bullet in the first three limbs, until it disintegrated in the right arm.

Two equally unusual cases have been reported by the FBI in the United States. In one, the victim was hit in the wrist with a .22 bullet. This was of sufficiently small calibre to travel up a vein and into his heart, where it killed him.

In the other case, a gunman held up a bank in Oklahoma. There were three people in the bank: one, the bank teller, had been at high school with him, and recognized him. To eliminate all the witnesses, the gunman forced them out behind the bank, made them kneel, and shot them one by one with a .357 Magnum. When it came to the bank teller's turn, the bullet entered her skull, travelled round inside her head, and exited from her forehead. She fell unconscious, and the gunman was sure that she was dead; but her brain was undamaged. In due course she recovered and testified against the gunman in court.

High velocity bullets can travel long distances from the scene of the shooting. When President Reagan was shot by John Hinckley, in 1981, FBI agents knew that six bullets had been fired. The wounded victims accounted for four, and the fifth had hit the right rear window of the limousine. Where was the sixth? The agents searched everywhere around the scene – and after sweeping up and examining every piece of street debris, they found it. It had struck an upstairs window across the street, and disintegrated. Fragments of the bullet were found below the window, but there was only a small hole in the glass.

Sometimes it is possible to determine a bullet's trajectory. A shot was fired through a window and into the wall of the Israeli consulate in Washington DC, and at first it was thought to be a terrorist attack. FBI experts aimed a laser beam from the point of impact on the wall, and through the hole in the window. The beam passed between other nearby buildings, and indicated an open area several blocks away. There, they discovered, a security guard had been chasing a handbag snatcher, and had aimed a shot at him. The marks on the bullet matched with test bullets fired from the guard's gun.

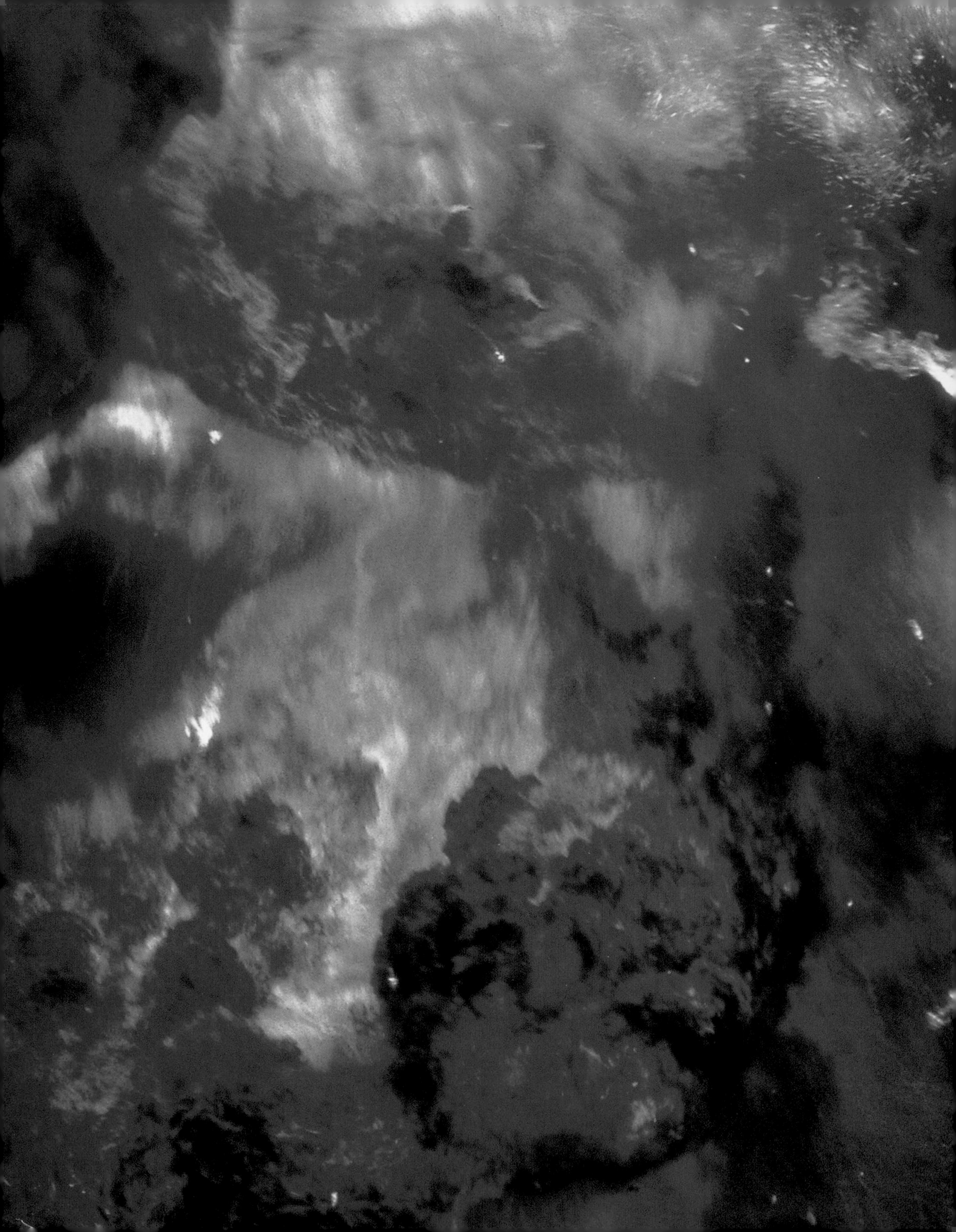

Fire and Destruction

A pillar of fire, smoke and burning debris is thrown into the air as the result of an explosion. The deep orange colour of the flames, and the thick black smoke, suggest the burning of a carbon-rich material such as oil.

EXPLOSIONS AND FIRES ARE CLOSELY RELATED: the chemical processes are very similar, and an explosion is often followed by fire, or a fire by explosion. Examining and identifying victims can present great difficulties to the pathologist and other examiners, while the investigation of the actual event requires the services of specialist experts.

EXPLOSIVES

Most people carry with them every day, or keep in a prominent place in their homes, a packet of explosive devices. These are matches, and they are freely on sale at every street corner. Nonetheless, they embody all the essential features of the most destructive explosive.

The first manufactured explosive was gunpowder, and its composition exemplifies all the requirements for an explosive. It is a mixture, principally of potassium nitrate (saltpetre) and charcoal, together with some sulphur. The potassium nitrate is a very rich source of oxygen, which combines with the carbon of the charcoal to form the gas carbon dioxide. In the open air, gunpowder burns quickly, but safely; however, if it is confined in a rigid container, the rapidly expanding gas produces an explosion.

Most modern explosives rely upon the same principle: the

What is an explosive?

- It may be a single substance or a mixture of substances.
- This substance or mixture is in a chemical state that is only temporarily stable.
- The outcome of upsetting this stability is the sudden release of a large amount of energy in the form of a hot, very rapidly expanding gas. Studies of explosions have shown that detonation waves can reach temperatures as high as 9032°F (5000°C), pressures as great as 1200 tons per square inch (200,000 kilograms per square centimetre), and speeds up to 18,000 miles per hour (8000 metres per second).

combination of carbon with oxygen. The more compact the form in which the oxygen source and the carbon can be packaged, the greater the force of the explosion will be. It is important that as much gas should be produced, and as little solid left, as possible. In this respect, gunpowder is particularly inefficient, as more than 50 percent of the products of explosion are solids. Many other explosives also leave a significant amount of residue.

Many millions of explosions take place on our streets every minute, inside the internal combustion engines of passing cars, trucks, buses and motorcycles. In each cylinder a mixture of atmospheric oxygen and fuel is compressed and then exploded. Explosions of domestic gas are essentially the same mechanism, the inflammable fuel being diluted with air in an enclosed space to form an explosive mixture.

Most people probably don't give a second thought to the simple match, but this familiar, domestic item possesses all the features of the most deadly explosive.

The oxygen source of an explosive may be a separate component of a mixture – as in gunpowder – or part of the molecule of a chemical compound that also contains carbon. In mixtures, the most favoured oxygen sources are nitrates and chlorates. Some of these are commonly available agricultural chemicals, fertilizers or weed killers, and have been used in many bomb attacks by terrorist organizations. The bomb built by Timothy McVeigh and Terry Nichols to destroy the Alfred P. Murrah Federal Building in Oklahoma City, on April 19, 1995, was of this type.

As organic chemistry developed in the 19th century, substances were synthesized that contained both carbon and nitrate groups in a single molecule. Among the earliest were nitroglycerine, trinitrotoluene (TNT) and picric acid (trinitrophenol). More recent explosives are tetryl (trinitrophenyl-methyl-nitramine), PETN (pentaerythritol tetranitrate, the principal explosive ingredient of Semtex) and RDX or Cyclonite (cyclotrimethylene-trinitramine).

Since these explosives are designed as unstable compounds, there is always the danger that they will spontaneously explode. Therefore, a product must be manufactured in a form that is relatively safe to handle. Nitroglycerine, for instance, is a liquid that explodes if it is dropped, or even shaken; Alfred Nobel's first dynamite, which he invented in 1866, consisted of nitroglycerine absorbed on to clay, making it relatively insensitive to shock. Nowadays, a variety of inert substances are used to "dilute" explosives, according to the use for which they are intended.

An explosion can be initiated by shock, friction (as in a match), flame or electrical discharge. What starts the explosion is a rise of temperature at a point in the explosive charge: once the reaction has begun it generates its own heat and spreads outward in a spherical shock wave, so rapidly that it is all over in a few millionths of a second.

Because a manufactured explosive must be safe to handle, it can be difficult to initiate an explosion; this is where the detonator,

Alfred Nobel, the Swedish explosives manufacturer who invented dynamite in 1866.

The manufacture of nitroglycerine involves the gradual mixing of concentrated nitric acid with glycerol. The temperature of the reaction must not be allowed to rise above 48° F (10° C). In the early days, the slow mixing was controlled manually, and the worker responsible was perched on a one-legged stool – to ensure that he did not fall asleep during the operation!

When traces of an explosive material are found on the hands of a suspect, it is usually taken in court as evidence that the person has handled that material with criminal intent. However, there is at least one other possible explanation. For many years doctors have prescribed nitroglycerine, highly diluted as a solution in alcohol or as tablets, for sufferers from angina pectoris. More recently other explosive organic nitrates, particularly PETN, have come into use. Given the high sensitivity of the tests for these substances, it is possible that traces could be found on the hand of someone who had recently taken tablets, quite legitimately, for a heart condition. It does not appear, however, that this possibility has ever been raised in court.

or "primary" explosive, comes in. It generates an intense shock, and so a very sharp rise in local temperature. This is a different type of explosive, generally being an unstable compound of a heavy metal with nitrogen, the explosion being caused by the release of the nitrogen. Mercury fulminate and lead azide are typical examples.

Investigating an Explosion

Criminal use of explosives falls into two categories: breaking and entering, whether through walls, roofs, floors or doors, or by attacking the locks of safes and strong-rooms; and attacks against persons or property, from either personal or political motives. An explosion investigator may be required also in cases of accident or suicide.

A forensic scientist examining materials, found at the scene of an explosion, for explosives residues.

As with any other forensic investigation, the scene must be approached with care. Debris which can be important as evidence may be scattered over a very wide area. A useful rule of thumb is to estimate the distance from the centre of the explosion to the furthest piece of debris, then seal off an area with a radius 50 percent greater than this. At this stage, photographs of the whole area should also be taken.

It is possible to establish the centre of explosion by closely examining the damage and calculating the direction of the shock wave. Long metal objects, such as piping, railings, window frames, furniture and storage racks – even long nails, screws or bolts – bend away from the direction of the blast. Metal sheets and plates, such as the doors of domestic equipment, are

An injured girl wearing an oxygen mask is attended to, 30 minutes after the Oklahoma blast.

The explosion that ripped through the Alfred P. Murrah Federal Building (left) in Oklahoma City, on April 19, 1995, was of a very simple type, employing a readily available weedkiller.

CRIME FILE:

Pan Am Flight 103A

Explosive experts were faced with the examination of some 4 million fragments from the destroyed airliner. They established the source of the explosion, a radio-cassette player packed with Semtex, and fitted with an electronic digital timer. This, and other evidence, led to two Libyans being charged with the massacre.

On December 21, 1988, 109 passengers left Germany's Frankfurt airport aboard Pan Am flight 103A, on the first stage of their journey to spend Christmas in the United States. At Heathrow airport in London, England, they had to wait six hours for their connection, aboard the Pan Am 747 *Maid of the Seas*, flight 103. Meanwhile their baggage, packed into a metal container, was transferred. At 6.04 PM, flight 103 was cleared for take-off. At 7.05 PM, as the aircraft passed over southern Scotland, and was cleared for its transatlantic crossing, it disappeared from the traffic control radar screens.

The aircraft had been ripped apart by an explosion and, in the high wind blowing in the upper atmosphere, the debris was scattered over a wide area, some pieces being eventually discovered over 100 miles away. The jet's No. 3 engine fell on the town of Lockerbie below, creating a crater 15 feet (4.5 metres)

The Pan American Maid of the Seas *exploded high above the small Scottish town of Lockerbie, raining down debris over an area of more than 800 square miles (2000 sq km).*

An investigator searches among the pieces of the Maid of the Seas. *Altogether, some four million fragments were recovered, some from over 100 miles (160 km) away.*

deep, while one wing produced a 30-foot (9-metre) crater, demolishing 2 houses and throwing up over 1500 tons (1524 tonnes) of soil and rock. All 259 people aboard the aircraft perished, together with 11 inhabitants of Lockerbie. The leader of one of the medical teams rushed to the scene reported that many of the bodies were reduced to fragments: "the pieces had just rained down".

About 4 million fragments of debris were recovered from 845 square miles (2188 square kilometres) of Scottish countryside, and painstakingly laid out at the Army Central Ammunition Depot, a short distance from Lockerbie. They revealed that the explosion had taken place in a metal container in the forward cargo bay on the lower left side of the fuselage. Trapped in the crumpled metal, an accident inspector found a tiny piece of printed circuit board. This was identified as part of a Toshiba radio-cassette player that had been packed with 14 ounces (397 grams) of Semtex, and placed in a brown Samsonite suitcase. The container was one that had been transferred from the Frankfurt flight at Heathrow.

The investigation was move to the Royal Armament Research & Development Establishment (RARDE). There, after weeks of examination of the debris, the discovery of more tiny fragments confirmed that the Toshiba player had been the bomb. Another fragment was found in the scraps of a shirt, and identified as part of an electronic digital timer manufactured in Zurich. Only 20 of the devices had been made, in 1985, to a special order from the Libyan government. In Senegal in February 1988, ten of these had been found in the possession of two Libyans – another turned up in the wreckage of a French aircraft that exploded over Niger in September 1989.

Other scraps of clothing were identified. One, a blue Babygro romper, bore the label "Malta Trading Company". The print-out listing the bags loaded aboard flight 103A at Frankfurt showed that one had been transferred from an Air Malta flight from Valletta – but no passenger from Malta had boarded the flight to London. Another clue was a fragment of trouser fabric found in the remains of the Samsonite suitcase.

Senior Scottish policeman DCI Harry Bell flew to Malta eight months after the disaster. He traced the trouser fabric to a shop, where the proprietor clearly remembered a man buying a quantity of clothing – including a blue Babygro – a month before the bombing. He described him as Libyan, aged about 50 and clean-shaven. It had been raining, and the man had also bought an umbrella. At RARDE, the investigators had found tiny fibres from the Babygro embedded in the fabric of the remains of an umbrella.

Almost three years after the bombing, United States and Scottish legal authorities released the names of two men who, they said, were responsible; and the photograph of one was recognized by the shop's proprietor. After a further eight years of legal wrangling with the Libyan government, the two were eventually put on trial in a Scottish court – which was unprecedentedly convened in Holland.

CASE CLOSED

CRIME FILE:

The Maguire Family

CASE CLOSED

Seven Irish men and women living in England were convicted of alleged involvement in IRA bombings. Fourteen years later, they were acquitted.

In England in 1976, six members of the Maguire family, and a family friend, were imprisoned for illegal possession of firearms. Their sentences were especially severe, due to the allegation in court that they were involved with the making of bombs for the IRA.

The prosecution presented evidence that the accused had traces of nitroglycerine on their hands, and the jury was assured that this test was conclusive. It was not until 1990 that the Court of Appeal overturned their convictions. Swabs taken from the hands and from rubber gloves said to belong to the Maguires had been analyzed by chromatography at the Royal Armament Research and Development Establishment, but the jury had not been told that other interpretations of the laboratory results were possible. They had also not been told that the results of subsequent tests – at least one of which was negative – had been withheld from the defence. Finally, it was alleged during the appeal hearing that the specimens could have been contaminated after arrival at the laboratory.

This case shows how explosives analysis must be carried out with great care and without any intention of reaching a foregone conclusion: every available test must be made to establish the presence of an explosive substance, and to identify it. It reminds us that even the presence of identifiable material on a suspect's hands is not necessarily evidence of being guilty of causing an explosion.

Left to right: Mrs. Anne Maguire, her husband, son Patrick, and Sean Smith outside the High Court, April 18, 1991 for the first day of the preliminary hearing into the "Maguire Seven" case.

"dished", empty metal containers also show the same effect, but full ones, such as water tanks and radiators, do not, because the liquids inside them are almost incompressible. Subsequently, in the laboratory, similar objects can be tested to determine the pressure that caused the damage. This can give a good idea of the nature of the explosive, and the quantity involved.

Other structural elements are normally blown outward by the blast, and objects of all kinds moved in that direction. Particularly indicative are any heaps of sand, dry earth or powder. Horizontal surfaces, such as worktops above cupboards, can prove very misleading. In the area of low pressure that follows the shock wave, they are likely to be lifted upward, suggesting – incorrectly – that the original explosion took place beneath them.

Fragments are very important. Dents – or, even more revealing, holes and scars – in vertical surfaces can help to pinpoint the centre of the explosion. Embedded fragments may determine whether the explosive was in some kind of container, and its nature. Pieces of the detonator, such as wires and the crimping cap, mechanical detonating devices, or small fragments of a timing device, may also be found. Explosion laboratories maintain a comprehensive collection of commercial products, from which it is often possible to identify the manufacturer and source of the explosive and its detonator.

If there are dead bodies or injured persons, they and their clothing must also be examined for traces of explosive and physical fragments. When an explosion is followed by fire, pathological examination is also essential (see pages 203–204).

When physical examination of the scene has been completed, chemical examination takes over. As nearly all explosives leave some solid residue, no matter how widely this has been scattered, traces can be found. Swabbing suspected areas with solvent may provide minute specimens of undestroyed explosive. There may also be traces of vaporized explosive, and several portable vapour detectors are available (see page 201).

In the laboratory, there are a number of "spot tests" using specific reagents, which identify many types of explosive. These are then followed by analysis of the specimens, using chromatography, and inorganic analysis for traces of mercury or lead in the residues. Finally, quantitative analysis of specimens helps to identify the particular commercial explosive employed.

An explosion ripped through an underground car park at the World Trade Center, New York, on February 26, 1993, killing 11 people and injuring more than 1000. FBI explosives experts were quickly on the scene. Traces of nitrate were found all over the area, and particularly on the remains of a van. The van had been rented in Jersey City by Mohammed Salameh, and more traces of nitrate were found on the rental agreement that he had signed. Follow-up investigation led the FBI to Nidal Ayyad and Mahmud Abouhalima – and their bomb factory.

Nidal Ayyad, one of the manufacturers of the bomb exploded in the underground car park of the New York World Trade Center on February 26, 1993, being led from court by a US marshal, after being denied bail.

Examination of Suspects

When the police apprehend people suspected of having caused an explosion, it is very valuable evidence if traces of explosive can be found about their person, in their clothing or possessions, or in a room they have occupied.

In recent years, with the development of trace detection techniques, particular importance has been placed upon the hands of the suspect. Even if gloves are worn while handling an explosive, there may still be some penetration to the skin.

Contamination of the hands can also occur due to transfer from another surface, such as a table on which explosive had been placed, or its wrapping material. Similarly, contamination can be transferred from the hand to another surface, such as the steering wheel or other controls of a car.

Disturbingly, transfer can also take place from the hand of one individual to another. It has been shown experimentally that a hand contaminated with an explosive could leave detectable amounts on a drinking glass, and that the glass could then transfer detectable amounts to the hand of another.

The suspect's hands are swabbed, first with a piece of dry surgical cotton wool, then with ether and finally with distilled water. After this, scrapings are taken from beneath the nails with wooden sticks, and all the specimens sealed in individual plastic bags. Analysis is then carried out by chromatography.

Fire Examination

Arson – deliberately setting fire to another person's property, or to one's own – usually has one of three motives: insurance fraud, revenge, or concealment of another crime, such as embezzlement, robbery or murder. There is also, of course, motiveless vandalism. Unlike most crimes, a case of possible arson involves no less than three independent inquiries: by fire officers, concerned with the cause of the fire; by the police, interested in discovering the perpetrator of any crime; and by the insurance investigators, who are naturally eager to detect any reason why insurance money should not be paid.

It is said – cynically perhaps, but unhappily with a fair degree of truth – that the incidence of arson is a clear measure of the health of the business community. When trade is bad, and businesses are failing, the number of arson cases increases.

When the fire investigation team arrives, their first priority is to ensure that the building is secured against further collapse: this, unfortunately, may result in some clues being obscured or destroyed. The fire investigators then look for the seat – the point of origin – of the fire. Since flames naturally burn upward, the lowest point at which fire occurred is likely to reveal important evidence.

The burnt-out premises of a discotheque in Gothenburg, Sweden, where 63 young people died in a midnight fire. Three young Iranians were later arrested, and admitted to the deliberate setting of the fire.

CRIME FILE:

The Unabomber

For nearly 20 years he was one of America's "Most Wanted". His home-made bombs bore characteristic "signatures", but nobody knew who he was. Only his letters to the press finally led to his identification.

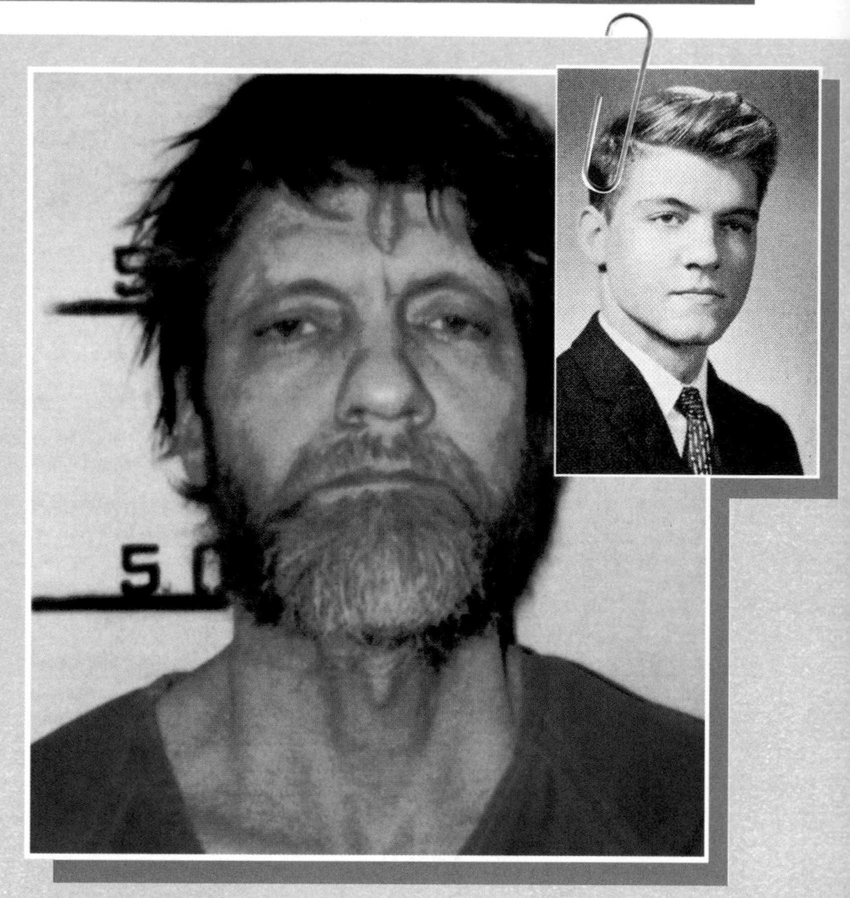

The face of Theodore Kaczynski, as it appeared in the Harvard University yearbook in 1962, and on his arrest in 1996.

The search for the man known as the "Unabomber" lasted nearly 20 years. The problem for FBI investigators was that his attacks followed no discernible pattern – university professors, aircraft, computer stores, an advertising executive, even a man who had lobbied against a wildlife protection programme. The Unabomber's campaign killed 3 people, and injured 29.

On the other hand, his bombs had a recognizable "signature". Every piece was carefully made by hand, or was something obtainable in any hardware store. The Unabomber made his own boxes, hinges and switches, cut wire to make nails, and filed away any identifying tool marks on screws. He carved pieces out of wood, and used old electric wiring and pieces of domestic plumbing. In many of the bombs, the letters "FC" were stamped on a piece of metal – possibly expressing his contempt for computers.

In November 1979, an explosive device caused a small fire in the baggage compartment of an American Airlines plane on a flight from Chicago to Washington. FBI agents found pieces of a home-made detonator, and soon learnt that similar devices had been used twice in the past 18 months.

The first bomb, a wooden box packed tight with the heads of matches, had been addressed to a professor at the Chicago, Illinois, campus of Northwestern University in May 1978. It was opened instead by a security officer, who was injured but survived. Over the years, the Unabomber's devices became more sophisticated. In December 1985 came the first fatality, a computer store owner named Hugh Scrutton.

In February 1987 the Unabomber was seen placing a bomb in the car park of a computer store in Salt Lake City, Utah. He was described as white, about 6 feet (2 metres) tall, aged around 40, with a medium build, ruddy complexion, and reddish-blond hair; he wore a hooded sweatshirt and aviator sunglasses. No other sightings were reported, and there were no bombings between 1987 and

1993, leading the FBI to suppose that he had been in some form of institution – possibly prison – during that time.

The FBI laboratory discovered a single clue in June 1993. A piece of paper found with the latest bomb bore an impression from writing on the sheet above; it read "Call Nathan R—. Wed. 7 PM". This scrap of evidence was, however, of little use on its own.

In December 1994, the Unabomber's most powerful device to date decapitated New Jersey advertising executive Thomas Mosser, as he opened a package addressed to his home. Then, in April 1995, he killed Gilbert Murray, lobbyist for the logging industry, in Sacramento, California.

In June 1995, a rambling 35,000-word manifesto claiming to be from the Unabomber was sent to the *New York Times* and the *Washington Post*, announcing that the bombings would cease if the letter was printed. Both newspapers complied, and there were no more attacks. Some time later, David Kaczynski of Chicago came across some papers written by his elder brother, Theodore – and realized that they contained phrases similar to those in the Unabomber's manifesto.

On April 3, 1996, Theodore Kaczynski was arrested at his mountain cabin in Montana. Now a skinny 55-year-old with a shaggy beard, he was sentenced to life imprisonment in January 1998.

The crude cabin in Montana where Kaczynski lived during the latter years of his career as the Unabomber.

FACT FILE

The investigating team first interview the firefighters on the type of fire, the observations they made while fighting it, and any suspicions they may have. Next, the investigators must discover the answers to a series of questions:

• What was the state of security of the premises? Were doors forced, or were there any other signs of break-in? Were there tool marks around locks, disconnected alarm systems or inoperative water sprinklers? Professional fire-raisers favour entering through the roof, where their point of entry is likely to be destroyed by the fire, but sometimes even obvious holes through ceilings or the walls of adjoining premises have been overlooked by the firefighters.

• Are there any material witnesses? Did anybody see anything suspicious? Was anyone seen leaving the premises?

• How many seats of fire are there? If the fire shows signs of having burnt from more than one source, its cause is at once suspicious.

• What was the quantity of combustible material present? Was it sufficient to cause the fire? Remarkably, even cases of arson do not always involve the use of accelerants such as petrol or other inflammable liquids.

• What was the source of the fire? Fire officers commonly divide fire into five classes: wood, paper and fabrics; hydrocarbons, usually inflammable liquids; electrical systems and equipment; combustible metals such as zinc and magnesium; and radioactive materials.

• Was the fire caused by gradual smouldering, or by a naked flame?

Wood floors and beams, for example, tend to carbonize in a checkerboard patern, the checks being smaller nearer the seat.

It is here that the debris must be sifted for any trace of some kind of timing device – most deliberate fires are started at night or on a quiet Sunday afternoon. One ingenious American arsonist connected a clock-radio to a light bulb. Many hours after he was out of town and had established a firm alibi, the clock-radio switched on the light bulb, which in turn ignited a fuse that led to a can of petrol. There was sufficient evidence left after the fire, however, to reveal the method he had used.

Sometimes an unusually heavy concentration of debris and ash can show where flammable material was piled together to start the fire. If a fuse of some kind – even twisted paper or fabric – was used, it may have left a discernible burnt trail.

If accelerants – petrol, paraffin, or other flammable liquids – have been used to start the fire, some will be absorbed by the charred wood or seep into cracks in the flooring, where it usually fails to burn from lack of oxygen. Investigators with a keen sense of smell can often detect traces, although it is impossible to distinguish alcohols, odourless paint thinners, and other non-aromatic liquids.

Fire following a car crash. If any part of the fuel feed is fractured, the electrical ignition system can rapidly initiate a fire, and an explosion will frequently follow, as the contents of the tank ignite.

Tests that are more sophisticated than the sense of smell have been developed for the detection of accelerants. One of the earliest used either one of two powders – Petrobst or Rhodakit – which were sprayed on to surfaces suspected of containing hydrocarbon inflammables, and changed colour on positive identification. These substances were not specific however, and their use has been abandoned. A handy portable instrument is now available; the British version is known as the "Gastec", and the American as the J-W Aromatic Hydrocarbon Indicator, or "snifter". Originally designed to detect inflammable gases in industrial plants, the instrument can register ten parts in a million. California's State Crime Laboratory reported its use in

CRIME FILE: Charles Schwartz

CASE CLOSED

He blew up his laboratory, and a charred body was found in the ruins. But was it his? Professor E.O. Heinrich successfully showed that the remains were those of another man.

America's first Professor of Criminology was Edward Oscar Heinrich of Berkeley, a chemist, who was dubbed the "Edison of crime detection" in the mid 1920s. He became famous throughout the United States in 1925, when he solved the mystery of the death of Charles Schwartz.

Schwartz, who also claimed to be a chemist, had announced that he had invented an artificial silk that was indistinguishable from the natural fibre, shortly before his laboratory was destroyed by a tremendous explosion on July 25, 1925. Mrs. Schwartz identified her husband's charred body in the ruins, and the police decided that the probable suspect was Gilbert Warren Barbe, a travelling missionary who was physically similar to Schwartz, and who had now disappeared.

A suspicious circumstance – the existence of a $200,000 life insurance policy, made payable to Mrs. Schwartz – prompted the police to consult Professor Heinrich. His first examination of the body showed that two teeth were missing from the dead man's jaw. A telephone call to Schwartz's dentist confirmed that he had extracted these teeth some time before, but Heinrich's examination indicated that they had been extracted very recently. He also noted that the eyes had been gouged out of the skull.

Heinrich next discovered that the dead man had been battered with a blunt instrument, and had been killed before the fire occurred. The fingertips, he found, had been burned with acid. When he obtained a photograph of Schwartz, he saw that his ear lobes were of a different shape from those of the corpse.

Schwartz's eyes, Heinrich was told, had been brown; Barbe's had been blue. He notified the police of his findings: the corpse was not that of Schwartz, but probably Barbe's. Schwartz was presumably in hiding somewhere, waiting for the inquiry to die down, before joining his wife and enjoying the fruits of the life insurance policy. The wanted man was discovered in a rooming-house in Oakland, California, but shot himself as the police hammered at his door.

Known as the "Edison of crime detection" and the "wizard of Berkeley", Dr. E.O. Heinrich was a leading independent forensic investigator during the 1920s.

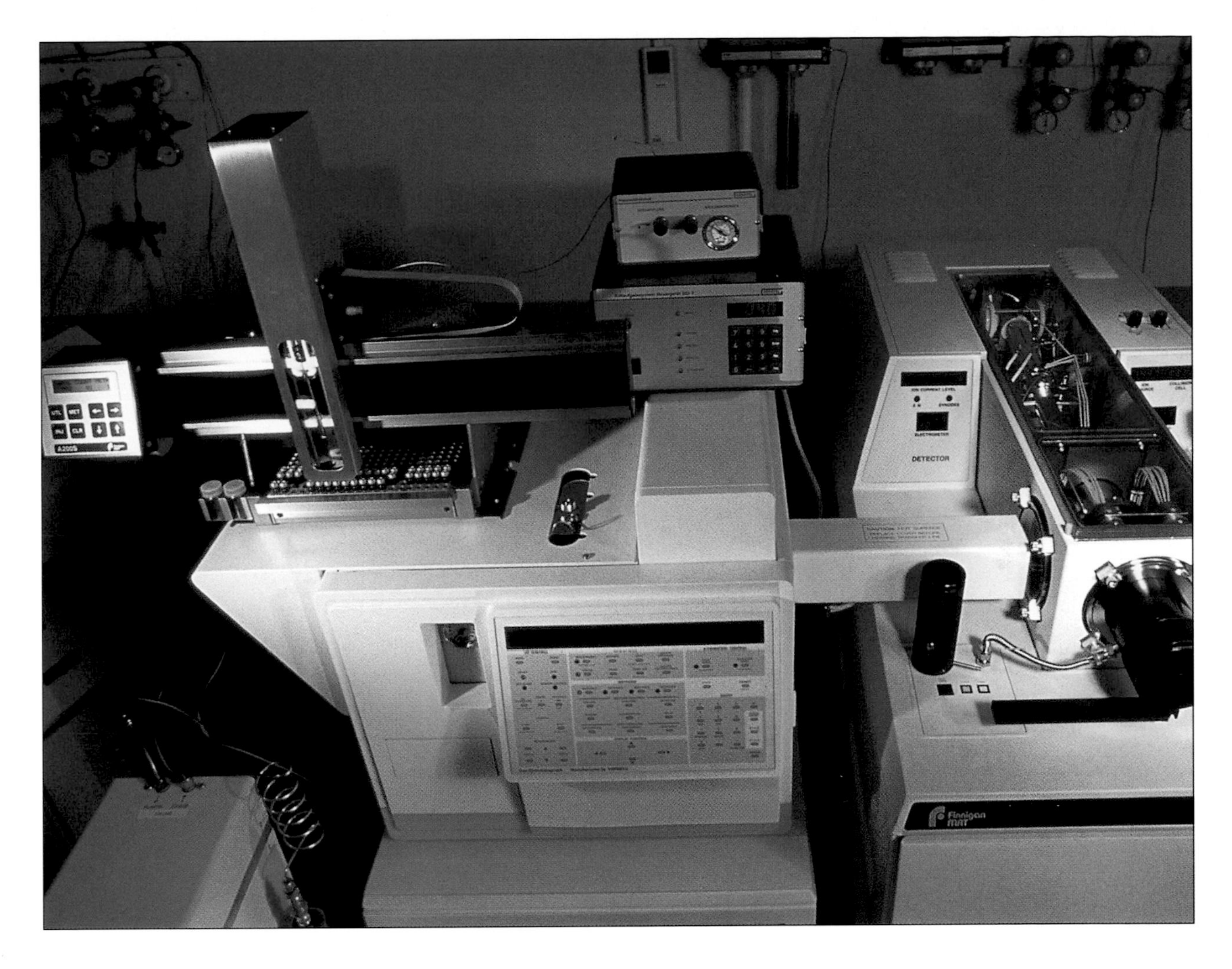

A gas chromatography machine (left) connected to a mass spectrometer (right). Gas chromatography is increasingly used in forensic investigation into the materials used to start a fire.

the case of a fire that gutted a farmhouse near Sonoma, in which two people died. The fire might have been dismissed as accidental, had the snifter not responded near one window, then registered strongly when lowered to the ground outside. It was clear that an arsonist had poured fuel through the window, and spilt some in the process.

Fire investigators now use gas chromatography to determine accurately the minute residues of accelerants used to ignite a fire. This work, however, requires the facilities of a fully equipped crime laboratory.

Dead Bodies

When a body is found in the debris of a fire, it is frequently lying with the fists clenched and arms raised in front of the body – the so-called "pugilist posture" – and the knees drawn up. Intense heat has caused the muscles to contract and stiffen almost immediately.

FACT FILE

Some fire authorities rely on the reported flame and smoke colour as an indication of the substances involved in a fire:

Material	Flame colour	Smoke colour
Acetone	Blue	Black
Benzene	Yellow to white	Grey to white
Cooking oil	Yellow	Brown
Fabric	Yellow to red	Grey to brown
Lacquer thinner	Yellow to red	Grey to brown
Lubricating oil	Yellow to white	Grey to brown
Naphtha	Pale yellow to white	Brown to black
Nitrocellulose	Red-brown to yellow	
Paper	Yellow to red	Grey to brown
Paraffin	Yellow	Black
Petrol	Yellow to white	Black
Phosphorus	White	White
Rubber		Black
Wood	Yellow to red	Grey to brown

It is essential to discover whether the victim was alive at the time of the fire, and died of asphyxiation – caused either by the immediate absence of oxygen consumed by the fire, or by smoke – or whether he or she was already dead, and the fire was started to conceal a murder. The pathologist takes samples of blood from the corpse; if they reveal the presence of carbon monoxide, it is certain that the victim was still breathing when the fire was started. Soot particles in the airways and lungs are another positive indication.

Despite extensive burns to the body, it is possible to discover whether any injuries were sustained before death occurred. When a living body is injured, the white blood cells (leucocytes) immediately migrate to the wound, producing a characteristic inflammation known as hyperaemia, and blistering. Liquid from these blisters can be tested for a positive protein reaction. Burns sustained after death are usually hard and yellow in colour, with few signs of blistering, and any liquid does not give the protein reaction.

Murderers who attempt to dispose of the evidence of their crime by setting a fire are seldom successful. The fire may be discovered and extinguished quickly; even if it burns intensely for a considerable time it is unlikely to reach a high enough temperature to destroy means of identification. Crematoria use a furnace that can attain temperatures up to 2732 °F (1500°C) but, even in these conditions, reduction of a body to ashes takes two to three hours. Even in cremated remains, inorganic poisons such as thallium can still be detected (see "With Poison Deadly").

Death by fire, whether accidental, malicious – or even suicidal – is horrific. Fortunately, homicide by burning is rare, as are cases in which fire is started to conceal a murder. Arson still remains a constant danger: as one British fire officer put it, "fire is a very available weapon".

CRIME FILE:

Kurt Tetzner

It seemed like a straightforward car accident, but the curious absence of carbon monoxide in the blood aroused suspicions in a case of insurance fraud and murder.

On November 27, 1929, a green Opel car was found in flames near Regensburg, Germany. It had apparently struck a milestone at the side of the road, and burst into flames, trapping the driver, whose burnt body was found inside. The owner was quickly discovered to be Leipzig businessman Kurt Erich Tetzner. His wife Emma identified remains of clothing as that of her husband.

The police were satisfied that this was a motoring accident, and released the body remains for burial. Insurance investigators were not so sure, as policies on Tetzner's life had been taken out only weeks previously. With difficulty, they obtained Emma Tetzner's permission for an autopsy, which was carried out by Professor Richard Köckel, from the Institute of Forensic Medicine at the university of Leipzig.

Köckel found that the remains comprised a badly charred trunk with spine attached, the base of the skull, the upper sections of both thighs, part of the right femur, and parts of the arms – nothing more, apart from a fragment of brain. He could find no soot in the windpipe, and blood samples revealed the absence of carbon monoxide. Perhaps Tetzner had been killed by the impact before the car burst into flames? Then Köckel began to compare the description of Tetzner with his anatomical findings on the body.

Tetzner was 26 years old, 5 feet 8 inches (1.7 metres) tall, broad-shouldered and heavily built. But the stage of fusion of the epiphysis in the upper arm (see Skull and Bones) indicated a person not much older than 20. Furthermore, the bones themselves were those of a lightly built youth.

Köckel reported his findings to the police. The remains were not Tetzner's, and the victim was dead before the fire was started. He suggested that the young man had been killed by a blow to the head, which was why the upper part of the skull had been removed, and the lower limbs were missing to prevent further identification.

Emma Tetzner was at once put under police surveillance, and a neighbour's telephone, which she often used, was tapped. On December 4, a call from a man in Strasbourg, France, was intercepted. The caller was told to try later, and in the evening of that day he was arrested in Strasbourg, and identified as Tetzner.

He made several varying confessions over the following months, but the general implication was that, with insurance fraud already in mind, he had picked up a hitchhiker, killed him and set fire to the car.

Tetzner was tried in Regensburg, found guilty, and hanged, in 1931.

CASE CLOSED

Fragments of Evidence

THE METICULOUS SCENE-OF-CRIME EXAMINER can spend many hours searching for trace evidence, often with a hand lens. He or she is looking for anything that does not naturally belong at the scene: usually something brought and left there by the criminal, although it may also be something associated with the victim. When a suspect is detained, every item of his or her clothing, possessions, and home and work environment, should be similarly examined.

Aspects of trace evidence detection are described in other chapters: prints of all kinds; blood, sweat, and other body fluids that can be analyzed by blood and DNA typing; bullets, spent cartridges and gunshot residues; fragments and residues at scenes of fire or explosion; and hairs and fibres. Traces of "foreign" material can also take many other forms, some of them extremely small.

A trace cannot be used as evidence unless it can be identified, and subsequently connected with the crime. Forensic laboratories maintain extensive databases of the physical characteristics and manufacturing details of glass, paint, paper, fabrics and other material, as well thousands of samples that can be used for comparison purposes. In the United States, for example, the National Automotive Paint File holds more than 40,000 original paint finish samples.

If this approach fails, or if the field of inquiry needs to be narrowed, manufacturers must be asked for further samples, as well as their records of manufacture, sales and distribution. All this can take many weeks – meanwhile a suspect remains at large, and the investigating detectives are increasingly impatient for laboratory results.

Car windscreens are constructed so that lethal fragments do not fly inward to injure or kill the driver, but the fractures spread out radially in a characteristic pattern.

Glass

Glass is an unusual substance. Apparently solid, it is in fact a liquid, cooled well below its point of solidification, and held between two highly stressed "skins" – a fact that accounts both for its transparency and its tendency to shatter into tiny pieces. These are properties of great importance to the forensic scientist.

When a criminal breaks into a building by smashing a pane of glass in a window or door, tiny pieces will almost certainly be found somewhere on his or her clothing. In court, defence lawyers have often claimed that, if somebody has broken a window by hitting it from the outside, the glass would all fall inward and so would not be found on the person who broke the glass. Police have known for a long time that this is not so. In 1967, Dr. D.F. Nelson of the then New Zealand Department of Scientific and Industrial Research finally resolved the argument. In collaboration with the Criminal Investigation Branch, Dr. Nelson had stroboscopic flash photographs taken of himself smashing panes of glass with a hammer. The photographs showed conclusively that, while some 70 percent of the glass fragments flew away from the striker, the remaining fragments flew toward Dr. Nelson, many of them hitting his clothes.

When a burglar smashes his hand through a pane of glass, fine fragments will be forced backward by the blow, and lodge in his clothing.

CRIME FILE:

Gerber's Foods

News of product tampering – a Federal offence – can create panic among consumers. But FBI analysis of glass fragments in baby foods proved this was not a campaign of contamination.

In 1988, newspapers all over the United States carried stories of mothers who were finding small fragments of glass in jars of Gerber's baby foods. There have been cases of light bulbs shattering over packaging lines, and a quantity of jars or tins being contaminated, but the fact that the glass occurred in different batches made this unlikely. Since product-tampering is a Federal offence, the jars were submitted to the FBI laboratory.

The examiner reported: "The glass in every jar was unique … mirror glass, headlight glass, light-bulb glass…In my opinion, it had been put in there by different consumers", presumably because they hoped that the company would compensate them.

CASE CLOSED

The same is true when a pane of glass is shattered by a bullet. Tiny particles can travel as much as 18 feet (5.4 metres) toward the person firing the gun.

A careful criminal may immediately brush all visible fragments from their clothing, and subsequently might even send it for dry-cleaning – but somewhere, caught among the fibres, specks of glass will remain.

Of course, most people have dropped and broken a drinking glass or something similar, and given no thought to tiny particles that may have lodged in their clothing. It is necessary for forensic scientists to prove that the glass is the same as that found at the scene.

There are several ways of establishing the identity of glass: one of the most elegant is by measuring its refractive index. Many different kinds of glass are manufactured for different purposes, and each kind – depending on the manufacturer and the type of glass – has a specific refractive index.

The refractive index of a glass sample can be measured from a particle with a diameter no greater than that of a human hair. The specimen is placed on a microscope slide and covered with a drop of fluid – usually a silicone oil, or one of a group of paraffin-like liquids known as Cargill fluids. The refractive

Everybody knows that a straight stick placed into water at an angle can appear to be bent. This is because water is a denser medium than air, causing light rays to travel slightly more slowly. Glass has the same slowing effect. The ratio, between the angle at which a ray of light hits the surface of the glass and the angle at which it passes through, is known as the refractive index. The refractive index of a vacuum (and, approximately, of air) is 1; glass, depending on its manufacture, usually has a refractive index between 1.5 and 1.7.

CRIME FILE:
Stephen Bradley

A young boy was abducted from near his home in Sydney and murdered, his body then wrapped in a rug. Traces of fungus, mortar, dog hairs, and seeds from a rare tree, led police to the home of the killer.

Stephen Bradley, aka Istvan Baranyay, escorted by Australian police following his arrest in Sri Lanka for the murder of eight-year-old Graeme Thorne.

Basil and Freda Thorne lived in a modest suburb of Sydney. On July 7, 1960, their eight-year-old son Graeme was abducted on his way to school. A man with a strong foreign accent telephoned his parents, demanding 25,000 Australian pounds for his safe return.

Over the next two days, police found some of Graeme's belongings dumped in distant parts of the city. There was, however, no further demand from his abductor.

On August 16, the boy's body was discovered about 10 miles (16 kilometres) from his home. He had been asphyxiated and clubbed to death. His body was wrapped in a rug, with one end tassel missing. His clothing carried traces of a pink, crusty substance; and mould had begun to grow on the boy's shoes and socks.

Hairs on the rug were identified as coming from three different people, and a dog, which Dr. Cameron Cramp of the State Medical Office stated was almost certainly a Pekingese.

The mould comprised four different types of fungi, and botanist Professor Neville White calculated that it had begun to grow about five weeks previously, indicating that Graeme had been killed shortly after he was kidnapped. The pink substance was identified as a type of mortar used in house facings.

Finally, over the course of a month, a quantity of leaves, seeds and twigs found on the boy's clothing was identified. Particularly interesting was a seed from a very rare species of cypress, which did not occur in the area where the body was found.

A mailman responded positively to a police broadcast asking if anyone knew of a pink house with this species of tree growing near. The house was empty: Stephen Bradley – a Hungarian immigrant whose real name was Istvan Baranyay – had left with his family on the day of the kidnapping, taking ship for England on September 26. Neighbours confirmed that he had owned a Pekingese dog. Inside the house, detectives found an old photograph of Bradley and his family seated at a picnic on the rug that had been wrapped around Graeme's body. They even found the missing tassel.

Learning that Bradley had sold his car on September 20, the police discovered it, still standing on the dealer's lot. In the boot were pink fragments identical to those discovered on the boy's body.

The ship on which the Bradleys were travelling was due to call at Colombo, Sri Lanka. Detectives flew there and arrested the murderer as soon as the ship docked. On March 29, 1961 he was found guilty and sentenced to life imprisonment.

CASE CLOSED

index of these liquids at room temperature is higher than that of any known glass, but decreases as the temperature rises. Identification is possible because the relationship between temperature and refractive index of the liquid is known.

The microscope stage (the platform on which the slide is placed for examination) is fitted with a heating element and a very sensitive temperature-measuring device. In modern laboratories the eyepiece of the microscope may be fitted with a video camera, so that the image can be enlarged even further and observed on a computer screen, and the signals from the various parts of the instrument fed into a computer for subsequent evaluation.

As long as the refractive index of the liquid is higher than that of the glass specimen, a faint halo – named the Becke line, from the scientist who first observed it – can be seen around the fragment. As the temperature of the liquid is gradually raised, its refractive index decreases. At the moment it becomes the same as that of the glass the Becke line vanishes.

The density of the glass – its weight per cubic centimetre – is also a valuable clue. It is not necessary to weigh the glass: if it neither sinks nor floats in a particular liquid, but remains suspended, then it has the same density as the liquid. By using a mixture of two liquids, one with a higher density than the glass and the other with a lower density, the forensic scientist can vary the proportions until the matching density is reached. If two different samples both remain suspended in the same mixture at the same temperature, their densities are the same.

Glass can be identified further by spectrography. When the glass is burnt in a high temperature carbon arc or laser beam, each element in the glass gives a characteristic coloration to the flame; the light is passed through a prism or similar device to produce a spectrum in which the specific wavelengths can be measured. This technique has a major drawback as evidence in a criminal trial, because the material sample is destroyed. Nonetheless, it is of particular value in the identification of coloured glasses.

In recent years, neutron activation has been used with success. As many as 70 different constituents can be identified in a speck of material no bigger than a full stop (period). The method is particularly valuable in that it can detect minute traces of elements not revealed by spectrography.

The identification of tiny specks of glass is important in establishing that a suspect was present at the scene of the crime. Larger fragments are of equal significance to the forensic scientist in such cases as automobile accidents and hit-and-run incidents. It takes great skill and experience to re-assemble the pieces of, for instance, a smashed car headlight, in the hope that its type, and the make of car from which it comes, can be ascertained.

Glass is a super-cooled liquid, and when the edges of the fragments are examined under a microscope, one can see that they have scooped-out ("conchoidal") fractures on the opposite side to that on which the breaking force was exerted. These fractures are of great help in fitting broken flakes together.

Finally, there are cases in which a pane of glass has been punctured by a shot or other missile, but has remained substantially intact. The type of fracture can reveal a great deal about what caused it.

Cone fractures usually occur when a high-velocity shot is fired through plate glass. The entry hole is smaller than the exit hole, and in this case all – or nearly all – the fragments fall on the exit side. Where the shot is of relatively low velocity (perhaps coming from a considerable distance), or a larger missile such as a stone has been used, the glass first bends before breaking, and a radial fracture, with lines radiating out in a starred pattern from the hole, is produced.

A blow with a pointed implement, such as the sharp end of a hammer, also results in radial fracture lines. In addition, it creates a "cobweb" of circular arcs round the centre point. Because these fractures happen on the far side of the glass, they can produce the tiny fragments that are sprayed back toward the person who has made the blow.

Where more than one missile penetrates the glass, it is possible to tell the order in which the holes have been made. The first missile creates radial fractures, as does the next; but in this case the radial fractures, if they meet a radial fracture caused by the first missile, usually terminate at that fracture line.

PAINT

Identification of paint fragments relies upon many of the same techniques that are used for glass. Most cases involve chips from the paintwork of cars; there may be eight or more layers applied during manufacture, and microscopic examination often reveals the undercoats along the edge of the chip. After the top colour has been identified by manufacturers' samples, these undercoats may make it possible to narrow down the inquiry to a single model of car, the plant from which it came, and even the period during which it was painted and subsequently delivered to dealers. Confirmation is obtained, if necessary, by spectrometry and gas chromatography.

In cases of hit-and-run automobile accidents, fragments of glass from the headlamp, found at the scene, can be identified, and frequently pinpoint the model of car responsible.

CRIME FILE:

Danny Rosenthal

The psychotic sadist, who carried out "experiments" on live chickens, murdered his mother and father. Examination of a hacksaw blade proved that it had been used to dismember the body of his father.

Danny Rosenthal was a 27-year-old American, who lived alone in a bungalow in Southampton, southern England. His parents were wealthy: his mother Leah lived in Israel, and his father Milton had a luxury apartment in Paris.

Late in 1981, both parents were reported missing. His mother had last been seen on a visit to her son's home, and the police obtained a search warrant. They found that Rosenthal – believed to be schizophrenic – had a "laboratory" in the bungalow, in which he had carried out bizarre experiments on live chickens. Among the birds' bloodstains that covered the floor, forensic scientists from the Central Forensic Research Establishment at Aldermaston found traces of human blood, and a hacksaw. Although its blade was missing, Dr. Mike Sayce discovered fragments of bone and human blood caught around its locking nut.

Meanwhile, French police had unearthed the dismembered remains, lacking head and hands, of a man whose characteristics matched those of Milton Rosenthal. In his apartment they found tiny fragments of bone, blood traces, and a hacksaw blade.

Together with French pathologist Professor Michel Durignon, Dr. Sayce compared marks on the bones of the dismembered corpse with the hacksaw blade. He cut through a block of wax, dusted the cuts with carbon powder, and obtained a match of tooth marks.

The hacksaw blade was proved to be one that Rosenthal had bought some weeks before. This evidence contributed to his conviction for murder in June 1982. Leah Rosenthal's body was never found.

CASE CLOSED

Tools

Many kinds of crime require the use of tools at some stage: burglary and safe-breaking, bomb-making, forgery – even product-tampering and the dismemberment of bodies. And every tool leaves its mark.

A burglar or thief generally uses a tool of some kind to effect entry, and the forensic examiner classifies the marks left as impressions, scratches, or cuts.

Impressions are found in such locations as wooden or aluminium window and door frames, where the metal of some kind of lever – the end of a hammer, a chisel or screwdriver, or a crowbar – is harder than the material of the frame. If a tool has been found in the possession of a suspect, it is often possible to match it with its impression.

Scratches on wood, metal or paintwork are caused by a knife or similar implement. Again they can frequently be matched to a tool found in a suspect's possession. Cut marks are left by such tools as knives, saws, and wire cutters; this category can also include the hardened drill bits used to penetrate locks or the

CRIME FILE:

Mark Hofmann

A dealer in rare books and papers, he made a small fortune in faking documents from Mormon history. But his final forgery revealed his ignorance of printing technology, and led to his downfall.

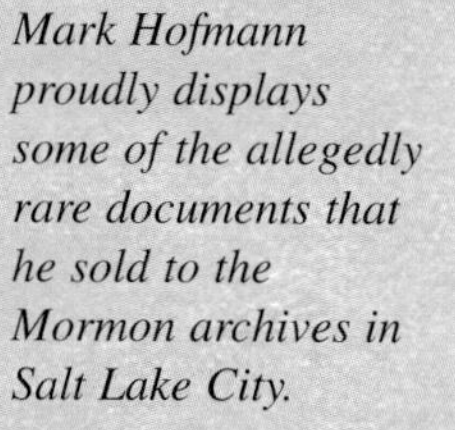

Mark Hofmann proudly displays some of the allegedly rare documents that he sold to the Mormon archives in Salt Lake City.

Mark Hofmann of Salt Lake City, Utah, dealt in rare books and documents. In 1985, he decided to forge the most famous lost document in American history. This was *The Oath of a Freeman*, a single sheet of paper little bigger than a postcard, printed in Massachusetts in 1639; its text was known, but all copies had disappeared. Another work from the same printer, however, the *Bay Psalm Book* of 1640, was easily available in a modern facsimile.

Hofmann photocopied pages from the *Bay Psalm Book*, cutting up and pasting down each character to assemble the text of the *Oath*, and had a printing plate made from it. He used this to print a piece of paper from a book of approximately the same date. To make ink, he burned part of the leather binding of another 17th-century book, to ensure that any attempt at carbon dating would establish its age. He allowed mould to grow on the paper to produce typical "foxing", and finally exposed it to ozone to oxidize and fade the ink. Then he offered the document to the Library of Congress for a million dollars.

Experts on ancient documents were unable to decide whether the *Oath* was genuine of not. It was sent to the cyclotron at the University of California for neutron activation analysis, and they reported that the ink appeared to be very similar to that in the *Bay Psalm Book*. William Flynn, a document expert at Arizona State Crime Laboratory, was not so sure. Asked about other documents that Hofmann had sold to the Mormon Church, he successfully showed that 21 out of 79 were suspect.

County Attorney Theodore Cannon finally pointed out that the *Oath* must be a forgery – without recourse to a cyclotron, carbon-dating, or even a microscope. He had spent 17 years as a letterpress printer, and knew all about type. Each metal letter is cast on a "body" that is greater in length than the distance between the top of the highest character (the ascender) and the bottom of the lowest character (the descender). The descenders in one line of type could therefore not come closer to the ascenders in the next line than the distance determined by the body. In several places in Hofmann's forgery it could be seen immediately with the naked eye that this was not the case: his cut-and-paste text did not retain the equivalent distances between ascenders and descenders.

There was also the matter of the border round the text. Once the individual type had been assembled, it would have been held in place by long strips of metal, before the border was set around it. In Hofmann's paste-up there was no need for this, and so there was insufficient white space between the text and the border. It could never have been set in type.

CASE CLOSED

walls of safes. These marks generally provide only an approximate match with the tool, because its cutting edge rapidly wears or suffers damage. There are, however occasional remarkable successes (see page 213).

Even the most ordinary "tools of the trade", such as the crowbar being used above to break and enter, leave impressions that can be matched to the tool – if it is found – and incriminate the perpetrator.

Paper and Ink

Documents, whether handwritten or printed, are often valuable evidence in an investigation. Handwriting itself can be of considerable importance, both in providing a psychological profile of the criminal and in leading to his or her identification (see "The Guilty Party"); while the paper and ink can be subjected to several kinds of analysis.

Ink falls into four principal types. The earliest form consisted of carbon black mixed with water, and this is still the basis of "Indian" ink, as well as the cakes of ink used traditionally in China and Japan. The blue-black ink that has been used for centuries is made from an iron salt combined with gallic and tannic acids, which were originally obtained from oak-galls. Coloured inks contain synthetic dyes, and are often water soluble. Ballpoint ink is similar, but can also contain insoluble pigments. All these types of ink also contain gum arabic, glycol, as well as other additives intended to prevent them soaking through paper, and make them more permanent.

A sheet of paper can take and retain an indented impression not only of what is written or printed upon it, but also of what was written on the sheet or sheets above it, as in a pad or notebook. For example, a receipt that no longer existed helped to send William Henry Podmore to the gallows, for the murder of his employer Vivian Messiter in Southampton, southern England in 1929. It had been torn from a pad by Podmore in an attempt to remove all evidence of his presence, but his writing had left a clear indentation on the following receipt.

Viewed by oblique lighting, Podmore's receipt was clearly legible, and photographs were produced in court. A more recent technique is the electrostatic detection apparatus (ESDA). It is based on the fact that pressure on paper can

CRIME FILE:
John Magnuson

He sent a parcel bomb to his neighbour. Unfortunately for him, vital scraps of the wrapper were recovered from the explosion. Expert examiners identified his handwriting, and even the pen and ink that had been used to write the address.

On December 27, 1922, James Chapman, a 65-year-old resident of Marshfield, Wisconsin, received a parcel that he took to be a delayed Christmas gift. Unfortunately, it was nothing of the kind: as he unwrapped it, it exploded violently. Chapman lost a hand, and his wife Clementine was fatally injured.

Scraps of the wrapping paper were recovered at the scene, and when they were put together they revealed a handwritten address: "J.A. Chapman, R.1, Marsfilld, Wis." Expert John Tyrell of Milwaukee, who examined the evidence, thought at first that the writing, which was ill-formed, had been disguised. A more detailed analysis indicated that it was genuine, and probably the best that the writer could manage. The mis-spelling of "Marshfield" suggested that he was foreign and probably, Tyrell opined, Swedish.

There was only one Swede in the area, 44-year-old farmer John Magnuson, who had feuded bitterly with Chapman over plans to cut a drainage ditch across his land. On December 30 he was arrested and, unaware that fragments of the bomb's wrapping had survived, agreed to provide a sample of his handwriting. It matched. At Magnuson's trial, two leading handwriting experts – Albert S. Osborn of New York, and J. Fordyce Woods of Chicago – agreed with Tyrell's findings. Osborn found 14 points of similarity, and concluded that "by no coincidence could any two persons ever make the characteristic and peculiar repetitions as displayed on these documents".

A further telling piece of evidence was Tyrell's observation that the address had been written with a medium smooth-pointed nib on a fountain pen, using Carter's black ink with a small admixture of Sanford's blue-black. A search of Magnuson's house uncovered just such a pen, belonging to his daughter. She customarily used Sanford's blue-black ink, but had lent the pen to a schoolmate, who had refilled it with Carter's.

Magnuson was sentenced to life imprisonment on March 31, 1923.

CASE CLOSED

affect its electrical properties, increasing its capacity to hold an electrostatic charge. ESDA comprises a flat bed of porous metal, on which the paper is laid, with a thin film of the transparent plastic Mylar laid above it. Vacuum suction below the metal bed holds the "sandwich" tightly together, and an electrical discharge is then passed over it. The indented parts of the paper become electrostatically charged and, when a mixture of photocopier toner and tiny glass beads is poured over the Mylar, the toner is attracted to the charged image, which can then be read.

The ESDA technique must be used before the paper is examined for latent fingerprints, however, because solvents affect the changed property of the indented paper.

CRIME FILE:

The Hitler "Diaries"

A publishing sensation turned out to be a crude forgery, written on paper that had not been manufactured until nearly ten years after Hitler's death.

On February 18, 1981, executives at the West German publishing company Gruner & Jahn were offered three diaries, written in an almost indecipherable script, that were said to be those of Adolf Hitler. Journalist Gerd Heidemann reported that they came from a wealthy collector, brother of an East German general.

The diaries were inspected by three experts – Dr. Max Frei-Sulzer, former head of the Zurich police forensic department; Ordway Hilton, from South Carolina; and a handwriting analyst from the Rhineland-Pfalz police. After many weeks, they declared them to be genuine. At once, Gruner & Jahn began secret negotiations for worldwide publication.

The publishers, however, had also submitted samples to West German forensic experts, including Julius Grant. They concentrated their examinations on the material of the diaries. On May 6, 1983, hours before publication was due, the experts announced that the documents were crude forgeries. The paper, which was of very poor quality, had been treated with a brightener that had not been used before 1954. The bindings of the diaries also contained brightener, and threads contained polyester, a post-war synthetic. And some of the ink was no more than 12 months old.

Heidemann, it turned out, had deposited payments for the diaries in his own bank account. His source turned out to be, not a wealthy collector, but a small-time criminal, Konrad Kujau. Both were sentenced to terms of imprisonment. As for Gruner & Jahn, it was estimated that the swindle had cost them a total of more than 20 million Deutschmarks ($16 million).

CASE CLOSED

The electrostatic detection apparatus (ESDA) reveals the impression of what has been written on the sheet of paper above. The indented parts of the paper become electrostatically charged and, when a mixture of photocopier toner and tiny glass beads is poured over the covering Mylar sheet, the toner is attracted to the charged image.

Speaking Likeness

"I RECOGNIZED HER VOICE." Evidence of this kind has sometimes been admitted in a court of law, and it reflects the fact that each human voice has specific characteristics that can be identified by ear. But this identification is subjective: the hearer may well be mistaken. The evidence should be presented in material form to convince a jury, and the only way to do this is by recording. When, for example, a confession is tape-recorded, meticulous precautions have to be taken to establish that the tape could not have been "doctored" in any way. And, even then, technical shortcomings can distort the sound so that the voice cannot be positively identified. Forensic scientists had to wait until 1967 to establish that a technique of speech identification – easily demonstrated to a sceptical jury – was acceptable as evidence in court.

The possibility of electronic voice identification became of considerable value during World War II when it was important to be able to distinguish different speakers in German military communications. Scientists and engineers at the Bell Telephone laboratories in New Jersey began work on the problem, and among them was Lawrence Kersta. He continued his work after the war, and in 1963 he finally developed a method of recording the pitch, volume and resonance of the human voice as what he called a "spectrogram". He wrote:

Recent advances in the development of speech synthesizers have been of value in voice analysis. This waveform represents the word "baby".

> The chance that two individuals would have the same dynamic use of patterns for their articulators [lips, tongue, teeth, etc, that distinguish the different sounds of speech] would be remote. The claim for voice pattern

CRIME FILE:

Edward Lee King

CASE CLOSED

When a young man boasted of his exploits in the 1967 Watts riots, he little realized that the analysis of his recorded voice would lead to his conviction. It was a landmark case.

The 1965 riots in the Watts district of Los Angeles were among the most destructive in American history. Looting and arson were widespread, and the events were widely covered by newspaper and television reporters. One CBS television interviewer recorded an encounter with a young man who, keeping his back to the camera, boasted of his fire-raising activities.

Some time later, the police arrested 18-year-old Edward Lee King and, suspecting that he had been the youth, asked voiceprint expert Lawrence Kersta to compare the TV recording with tapes of King's voice. At a trial that lasted nearly two months, Kersta testified that the voiceprints were identical, and King was convicted of arson.

An important legal issue was raised when King appealed, on the grounds that giving a voice sample amounted to self-incrimination. Eventually the US Supreme Court ruled that the right of privilege against self-incrimination did not apply in this case.

August 13, 1965, and clouds of smoke blanket the Watts district of Los Angeles as the mob loots and burns in a three-day riot.

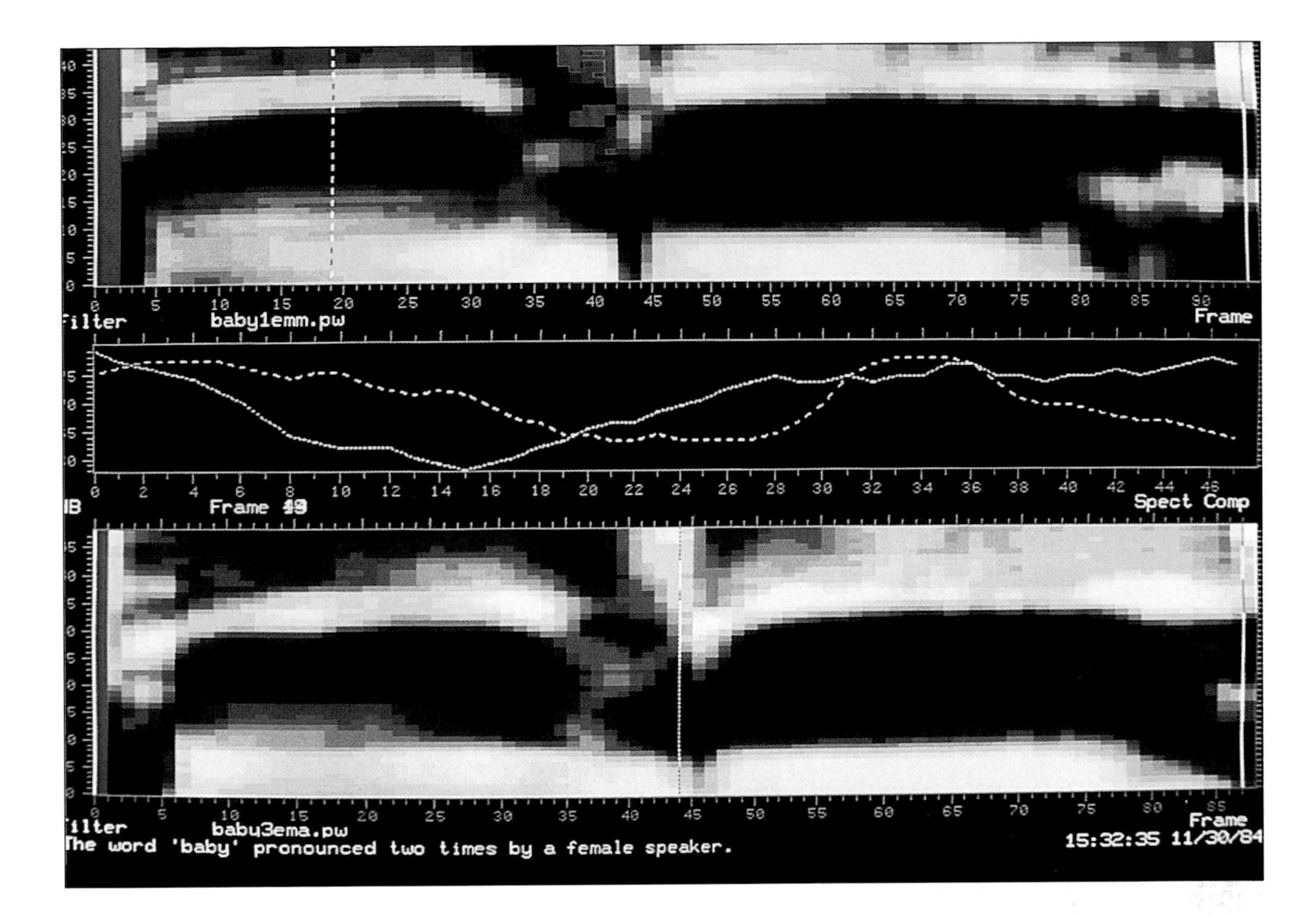

Computer graphics voiceprints of a female voice speaking the word "baby". Even though the level of the sound volume – represented by the two traces in the centre – varies greatly, the "spectograms' are very similar.

uniqueness, then, rests on the improbability that two speakers would have the same vocal cavity dimension, and articulator-use patterns, nearly identical enough to confound voiceprint identification methods.

Kersta's apparatus records a 2.5-second band of speech on magnetic tape, which is then scanned electronically. The scan can be displayed on a cathode ray screen, or recorded by a stylus on paper around a rotating drum. Two types of voiceprint can be obtained. One, which is usually produced as evidence in court, is the bar print: the horizontal scale represents the length of time of the recording, and the vertical scale represents the frequency of the sound. Loudness is represented by the density of the print. The other type is the contour print, which displays the more complex characteristics of the voice, and is suitable for filing on computer.

The most commonly used words in speech are "a, and, I, is, it, on, the, to, we, you". In establishing the reliability of voiceprinting, Kersta made 50,000 recordings of individual voices. Many sounded very similar, but differences

CRIME FILE:

Myra Hindley and Ian Brady

Sadistic killers of young children, the "Moors murderers" made tape recordings of some of their victims' last moments. And the sound of a radio broadcast helped pinpoint the time at which a 10-year-old girl died.

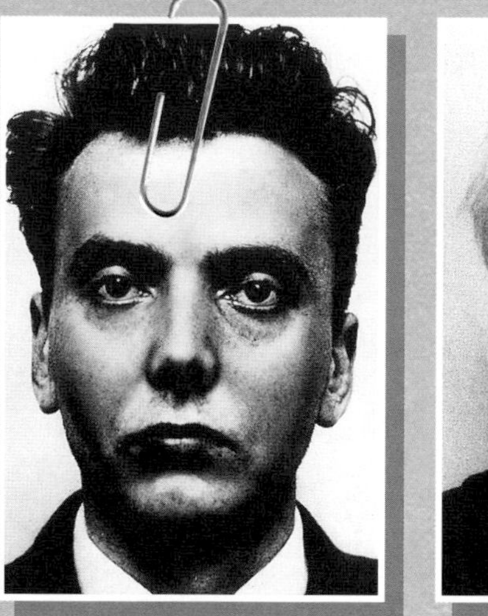

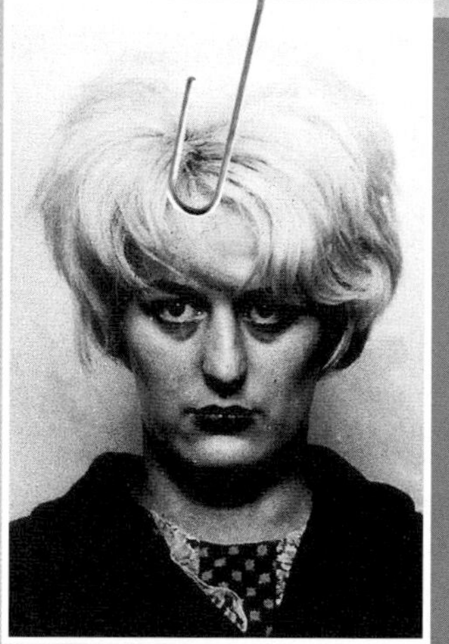

The "Moors Murderers" Ian Brady (left) and Myra Hindley (right).

One of the most devastating items of tape recording was produced as evidence in a British court in 1966, during the trial of Ian Brady and Myra Hindley for the murders of several young people. It bore the voice of 10-year-old Lesley Ann Downey, pleading with Brady as he forced her to strip for pornographic photographs, raped and strangled her. Lesley had been abducted on December 26, 1964, and her body, buried on a nearby moor, was not discovered until October 16, 1965. But in the background of the tape could be heard a radio broadcast from Radio Luxembourg, with the voice of singer Alma Cogan clearly detectable – providing a definite date for the recording.

English police search marshland on a wind-swept Yorkshire moor for the buried body of one of the victims of Ian Brady and Myra Hindley.

were clearly visible in a cathode ray display. He also made use of professional mimics, demonstrating that, although they were indistinguishable by ear, their voiceprints were markedly different.

Since 1967, voiceprint evidence has been occasionally admitted in the American courts, although it is still viewed with considerable scepticism in Europe. A watershed case was the trial of poacher Brian Hussong for the murder of Neil LaFeve in Wisconsin in 1971.

In the same year, Lawrence Kersta was asked to give his opinion in a very different case. Author Clifford Irving had offered the McGraw-Hill Book Company the manuscript of what he claimed to be the authorized biography of the eccentric millionaire Howard Hughes. To back up his claim, he produced letters he said had been written to him by Hughes, which were pronounced genuine by handwriting experts. But, after 15 years of reclusion,

Millionaire Howard Hughes, photographed around the time he spoke before a US Senate sub-committee. A recording of his voice, made at the hearing, provided the clinching evidence in the prosecution of Clifford Irving some 30 years later.

Clifford Irving, who very nearly succeeded in swindling publishers McGraw-Hill out of $650,000 with his forged Howard Hughes biography.

Howard Hughes finally broke his self-imposed silence to announce that Irving's so-called biography was in fact nothing less than a "totally fantastic fiction".

Hughes made his claim in a two-hour telephone call from his hideaway on Paradise Island in the Bahamas. The question was whether the voice on the telephone was that of Howard Hughes? Kersta examined tapes of the conversation, comparing them with a recording of a speech made thirty years earlier before a Senate sub-committee. He announced that the voice was undoubtedly that of Hughes, and declared: "We are as near to 100 percent positive as a scientist would ever allow himself to be." In June 1962

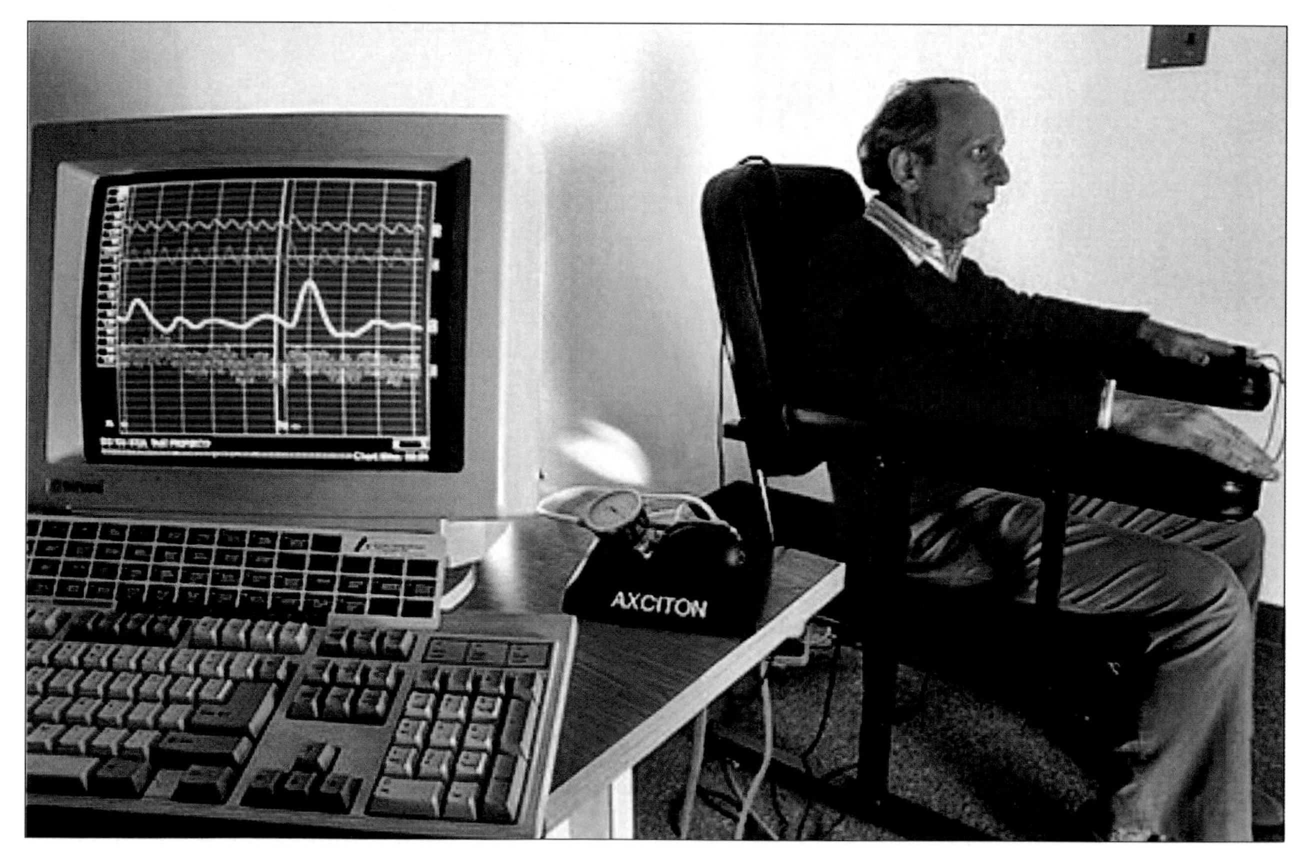

A man undergoing a polygraph lie detector test, which records changes in pulse, blood pressure, and breathing. The measurements are displayed on the monitor in the foreground.

CRIME FILE:

Brian Hussong

A well-known poacher, he shot and decapitated a game warden. The police could not find the suspect's gun, and it was the voice of his grandmother, identified from a wire-tap, that betrayed him.

Neil LaFeve was a game warden in the Sensiba Wildlife Area near Green Bay, Wisconsin. On September 24, 1971 he did not return from his work, and on the following day police discovered his decapitated body in a shallow grave, with his head buried close by. He had been shot several times.

The officer in charge of the investigation, Sgt Marvin Gerlikovski, suspected that this was probably a revenge killing, and ordered that all those previously arrested by LaFeve for poaching should be found and questioned. Those who were unable to furnish an alibi were asked to take a polygraph test, and all agreed except one – a notorious local poacher, Brian Hussong.

Gerlikovski applied for a court order permitting him to place a telephone tap on Hussong's house, and made recordings of all conversations on the line. One was with Hussong's 83-year-old grandmother, Agnes Hussong, who assured him that all his guns were well hidden. A brief search of her home uncovered the guns, which were sent to the state crime laboratory, and ballistics expert William Rathman quickly confirmed that one matched the .22 shells found near LaFeve's body.

At Hussong's trial, the grandmother denied that she knew anything about his hidden guns. However, Ernest Nash of the Michigan Voice Identification Unit gave testimony that voiceprints definitely identified Agnes Hussong on the recorded tapes, and demonstrated the differences from others of Hussong's relatives. Hussong was found guilty, and sentenced to life imprisonment.

CASE CLOSED

Irving was found guilty of forgery, and sentenced to a term of imprisonment.

More recently, voiceprint recognition has been employed by the USAF in security systems. An authorized person records a sequence of phrases into a computer memory. When he or she wishes to gain access to a restricted area, one or more of these phrases is compared by the computer before access is permitted.

Two new instruments have been developed along the lines of voice analysis. Both are intended to replace the polygraph in lie detection. Unlike the polygraph, these devices do not have to be attached to the suspect to detect changes in heart rate or perspiration, and it is claimed that they can produce accurate indications from live speech, over a telephone, or from tape recordings. The Psychological Stress Evaluator (PSE), it is claimed, will detect subaudible tremors in a suspect's voice whenever he or she is telling a lie. Although findings with these instruments are unlikely to be admitted as evidence in court, they can be valuable in indicating the direction in which further investigation should be directed.

The Guilty Party

ONE OF THE PERSISTENT PROBLEMS IN CRIMINAL INVESTIGATION is identifying the perpetrator of a crime. Witnesses at a scene can be mistaken, or disagree; or a suspect may find people who genuinely believe that they have seen him or her far from the location at the time. When there is no positive identification of the perpretator, the police must look for clues that indicate a likely suspect: all police forces maintain records of the *modus operandi* (MO) – the pattern of operation – of known criminals, and this helps to narrow down the search; while in recent years, ways of analyzing the criminal personality and psychology have resulted in an impressive number of successes. Finally, when a criminal is apprehended, and particularly when it comes to passing sentence in court, it is important that his or her identity is unequivocally established.

When it comes to the unequivocal identification of a criminal, the devil is truly in the detail. The perpetrator of a crime can be revealed by the tics and idiosyncracies of behaviour or – as seen here – by the unique characteristics of handwriting under microscopic examination.

In the 19th century, many criminologists believed that it was possible to identify a "criminal type". The leading figure among them was the Italian Cesare Lombroso, who published *L'Uomo Delinquente* (*Criminal Man*) in 1876. After studying nearly 7000 criminals, Lombroso came to the conclusion that their physical appearance was directly related to the type of crime they committed. Although his theories have now been largely discredited, they provided a stimulus to the science of anthropometry – the study of the variations in physical measurements of different human types.

The president of the Paris Anthropological Society at that time was Dr. Louis Adolphe Bertillon, who was involved in comparing and classifying the shape and size of the skulls of different races. His son, Alphonse, appeared to show little interest in this work, but when he obtained a post as a junior clerk in the records office of the Paris Prefecture of Police, he realized that his father's methods could be applied to the identification of known criminals. He recalled the statement of the Belgian statistician Lambert Quetelet, that no two people shared exactly the same set of physical measurements, and outlined a proposed identification system to his superiors. Between November 1882 and February 1883, Bertillon assembled a card file system of 1600 records, cross-referencing the measurements he made on arrested criminals. His technique soon became known as "bertillonage".

On February 20, 1883, a man calling himself Dupont was brought in. Bertillon took the man's measurements and, without much hope, began to go through his cards. Gradually his excitement mounted until, with a triumphant flourish, he drew out a single card. "You were arrested for stealing empty bottles on December 15 last year," he cried. "At that time you called yourself Martin."

The new card-file system received sensational publicity in the Paris newspapers. By the end of 1883 Bertillon had identified nearly 50 recidivists, and the following year he identified more than 300. Bertillonage was quickly adopted throughout France by police and prison authorities.

Cesare Lombroso, the leading figure among 19th-century criminologists who hoped to be able to identify the anthropological characteristics of the "criminal man".

Following this, Bertillon turned to photography. He established the practice – still in use today – of taking full-face and profile portraits. He also introduced what he called the *portrait parlé*, a system of accurately recording in writing the shape of facial features such as the nose, eyes, mouth and jaw. This, too, is still taught to trainee detectives, and is the basis of Identikit and other modern identification methods (see pages 234–235).

The development of fingerprinting (see "Finger of Suspicion") put an end to the adoption of bertillonage in any other country, but Bertillon himself stubbornly adhered to it.

Facial Recognition

The identity (ID) parade is a familiar feature of movie and television drama, and a regular part of many a police investigation. Now, however, psychologists – and many police forces – are beginning to question the validity, both scientific and legal, of the system.

It is essential that the suspect in the police lineup should not be indicated in any way to the witness. A number of other persons must therefore be found, among whom the suspect does not stand out in any way. Ideally, they should

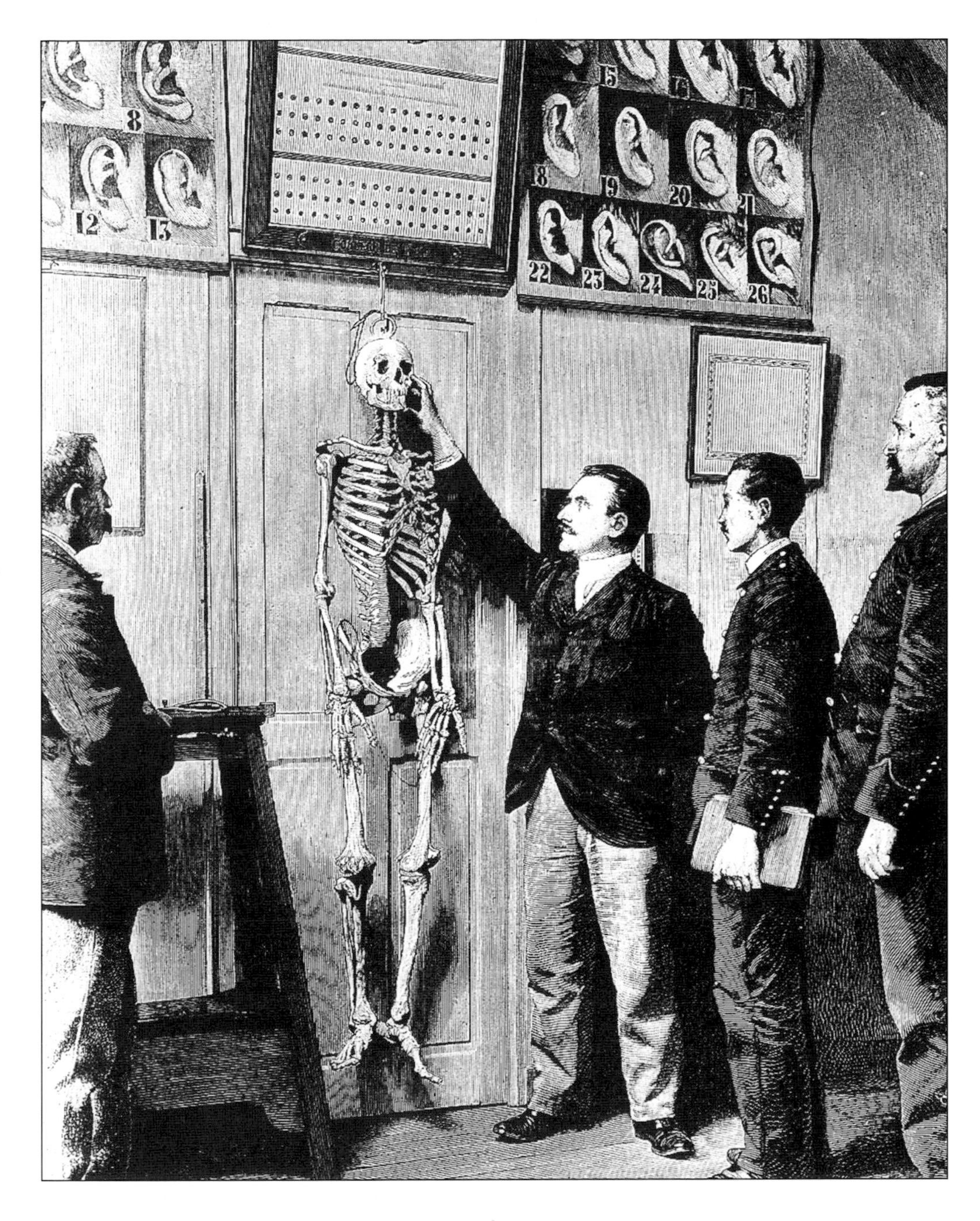

An engraving from 1899 depicting lessons in anthropometric measurements at Paris police headquarters, where Alphonse Bertillon developed his system of "bertillonage".

all be of the same physical type, in height, build, hair and facial colouring. In practice, people are often stopped on the street near police headquarters and asked if they are willing to take part in the parade. Since most are likely to have little time to spare, and since the witness is unwilling to wait long, the number assembled is unlikely to be sufficient. Frequently the necessary minimum is made up by off-duty officers hastily fetched from the canteen.

CRIME FILE:

James Hanratty

In one of the most infamous cases in British justice, Identikit pictures of the suspected rapist and murderer did not match the arrested man, but he was found guilty, and hanged. Later, a second suspect confessed to the crimes.

On the evening of August 22, 1961, Michael Gregsten, a married man, and his lover Valerie Storie, were sitting in his parked car beside a field near Slough, southern England. Suddenly, a man opened a back door of the car and climbed in, with a gun in his hand. He ordered Gregsten to drive for hours to a lay-by on the A6 at a spot named Deadman's Hill, where he shot him dead. He then raped and shot Storie, and drove away. By a miracle, she survived, although she remained permanently disabled. A loaded revolver, identified as the murder weapon, was later found on a London bus. The car was also found abandoned in London.

Valerie Storie gave a description of the murderer, and an Identikit portrait

James Hanratty, who was hanged on April 4, 1962 for the A6 murder in spite of serious doubts about his identification.

The spot on the A6 at Clophill, Bedfordshire, where Valerie Storie was found lying seriously injured beside Michael Gregsten, who had been shot through the head. The white canvas screen marks the place where his body was found.

was produced. It differed in nearly every detail from another Identikit picture, prepared from descriptions by three witnesses who had later seen a man driving Gregsten's car. The only common feature was the "deep-set brown eyes" that Storie positively remembered.

Meanwhile, police investigators had detained two suspects: James Hanratty, in whose hotel room detectives discovered shells from the murder weapon; and Peter Alphon, who had occupied the room on the following night. Hanratty did not look like either of the Identikit pictures: he did not have dark hair brushed back, and his eyes were pale blue. On the other hand, Alphon closely resembled Storie's description.

Then, for some reason that has never been explained, Storie changed her mind: she said that her attacker had "icy-blue" eyes, and that they were "saucer-shaped". She failed to identify Alphon in an ID parade, but three weeks later picked out Hanratty in a second line-up. He was immediately arrested, found guilty at trial, and hanged in April 1962.

Hanratty had insisted that he had been in northern England on the night of the murder, and witnesses later came forward to support his alibi. Furthermore Alphon, free from the fear of prosecution, made a series of statements that he had been hired "by an interested party" to break up the affair between Michael Gregsten and Valerie Storie. These events threw even further doubt on the validity of the Identikit identification, and public dissatisfaction with Hanratty's conviction has continued to the present day.

CASE CLOSED

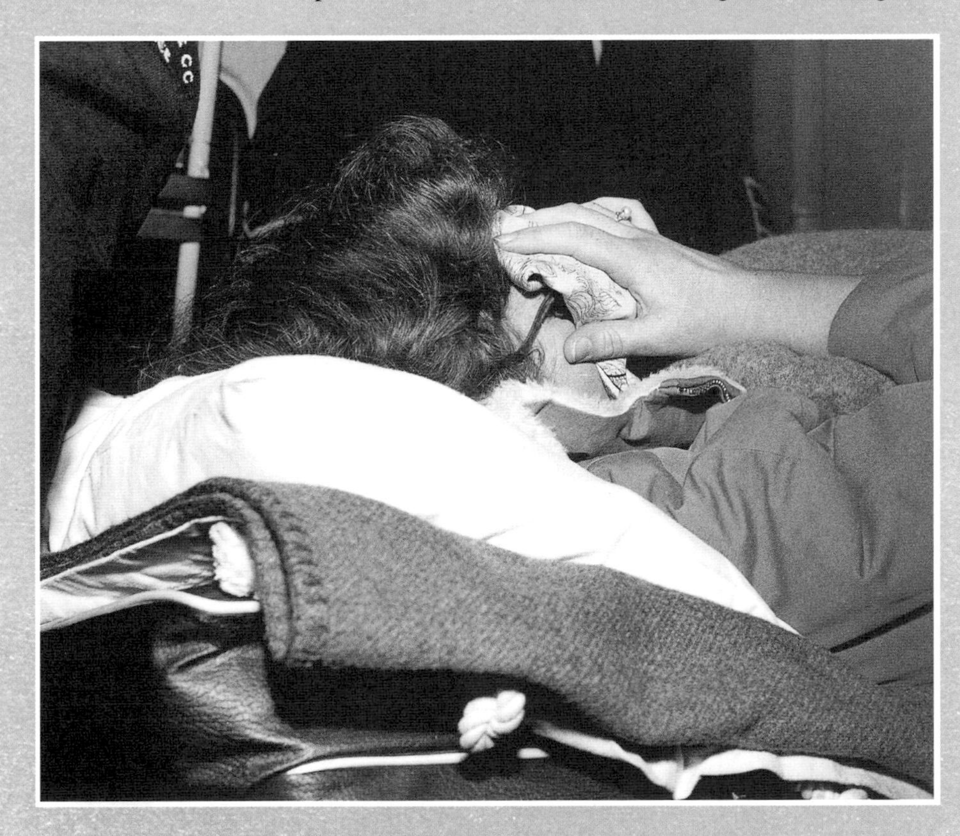

Valerie Storie covers her face as she is carried into an ambulance outside Bedford Assizes, where she gave testimony that the injuries she suffered after being shot left her paralyzed from the waist down.

The circumstances of the ID parade often make everyone who stands in it look guilty. In addition, the witness feels under considerable pressure to identify someone. Dr. Donald Thomson, lecturer in psychology at Monash University in Australia, has been fiercely critical of the identification process. One day, after he had made some outspoken comments about the methods of the New South Wales police on a television programme, he was brought in off the street and made to stand in an ID parade.

A woman had been assaulted in her home. She looked briefly at the row of men – and unhesitatingly identified Thomson as her attacker. "My first thought" he said, "was that the police were trying to scare me." Fortunately, he

FACT FILE

How do we distinguish one individual from another? At one time it was suggested that single brain cells might be imprinted with information to enable us to recognize any one object that we might come across. For instance, there would be "grandmother cell", which would recognize and identify one's grandmother. But, as Professor Whitman Richards of the Massachusetts Institute of Technology pointed out: "If you have...cells to react to every possible animal or thing that you might see, you're going to run out of cells pretty quickly."

One of Whitman's colleagues, Englishman David Marr, came up with a different theory. He proposed that the eye first fed the brain with a quick overall impression – like an artist's "lightning sketch", in which just a few lines delineate the object. Signals from cells in the eye would detect the contrast between light and dark, and would be sufficient to identify the general form of the object. Marr and his colleagues built one of the first scanning computers on this principle, and produced pictures very like artist's sketches. Today's security scanners can recognize a visitor's face, and compare it with authorized faces stored in its memory.

Marr suggested that the brain, having first identified the object from such a crude sketch, then gradually focused closer and closer on the important features, building up a complete detailed image to be stored in the memory.

The first features noticed by a witness are the hair, mouth and eyes: the colour, shape and length of the hair; the shape and attitude of the mouth; and the shape and colour of the eyes. Dark glasses, for instance, can so change the appearance of the face that it can be unrecognizable at first glance. Next comes the overall form of the face, a "sketch" such as Marr's computer produced. Only if the witness has sufficient time to "focus" on the details is it possible to identify the face completely, compare it with what is stored in the memory, and recognize it. In the case of a familiar face – a member of the family, a friend or someone very famous – this takes just a fraction of a second. It is not surprising that, faced with a suspect in an ID parade who approximately resembles someone they saw under conditions of stress, and feeling obliged to make an identification, witnesses frequently identify him or her as the perpetrator of the crime.

Early identity parades took none of the precautions that are now essential. Today, the possibility of intimidation of the eyewitness is avoided by placing him or her behind a one-way mirror, while in some jurisdictions, the witness views only one suspect at a time.

had an alibi: at the time of the assault he had been on air, broadcasting from the television studio. It then transpired that the woman's television had been on, tuned to the broadcast; in the stress of the attack the image of Thomson's face had somehow been superimposed, in her memory, on that of her attacker.

As Thomson said: "In the midst of trauma and shock, we can erase a lot of our capacity to remember. Afterwards…we thread associated memories together in order to reconstruct the incident in our minds."

This is not the only danger in the identification process. Prejudice can also play an important part in distorting what the witness thinks he or she has seen. Some years ago in England, the London Metropolitan Police organized a publicity campaign that featured a photograph of a street scene, and asked "What would you do?" Running toward the right of the picture was a black man in a short ski jacket and an open-necked shirt. Fast behind him came a uniformed police constable. The implication seemed clear: the black man was a criminal, being pursued by the constable – but the text accompanying the photograph pointed out that the black man was actually a plainclothes detective chasing a suspect, outside the frame of the picture, with the back-up assistance of the uniformed officer.

The police are becoming increasingly aware of what a dangerous role prejudice can play in eyewitness identification. Among the instructional aids at Hendon Police Training Centre in north London was a video showing a man with short spiky hair, and wearing a leather jacket, snatching a handbag from a middle-aged woman. After watching the video, trainees were asked the age of the bag-snatcher. Almost invariably they replied, confidently, that he was in his 20s – but he was in fact over 50. They had applied their memories of typical spiky-haired wearers of leather jackets to what they had seen, and assumed that the bag-snatcher must be the same age.

In the present climate, the police are particularly sensitive about wrongful conviction. On the other hand, they are still aware that eyewitness evidence can be invaluable. In some states in the United States, the police have already abandoned the traditional form of ID parade, in favour of "sequential identification", where the witness examines a single face at a time. Psychologists maintain that this enables the witness to concentrate more effectively on each face, and recent studies have suggested that the sequential method reduces the possibility of false identification by half.

IDENTIKIT, PHOTOFIT, VIDEO-FIT AND FACEIT

In the late 1940s Hugh McDonald, chief of the civilian division of the Los Angeles Police Department, was in Europe, attempting to track down crooks and conmen who were exploiting the confusion that followed World War II.

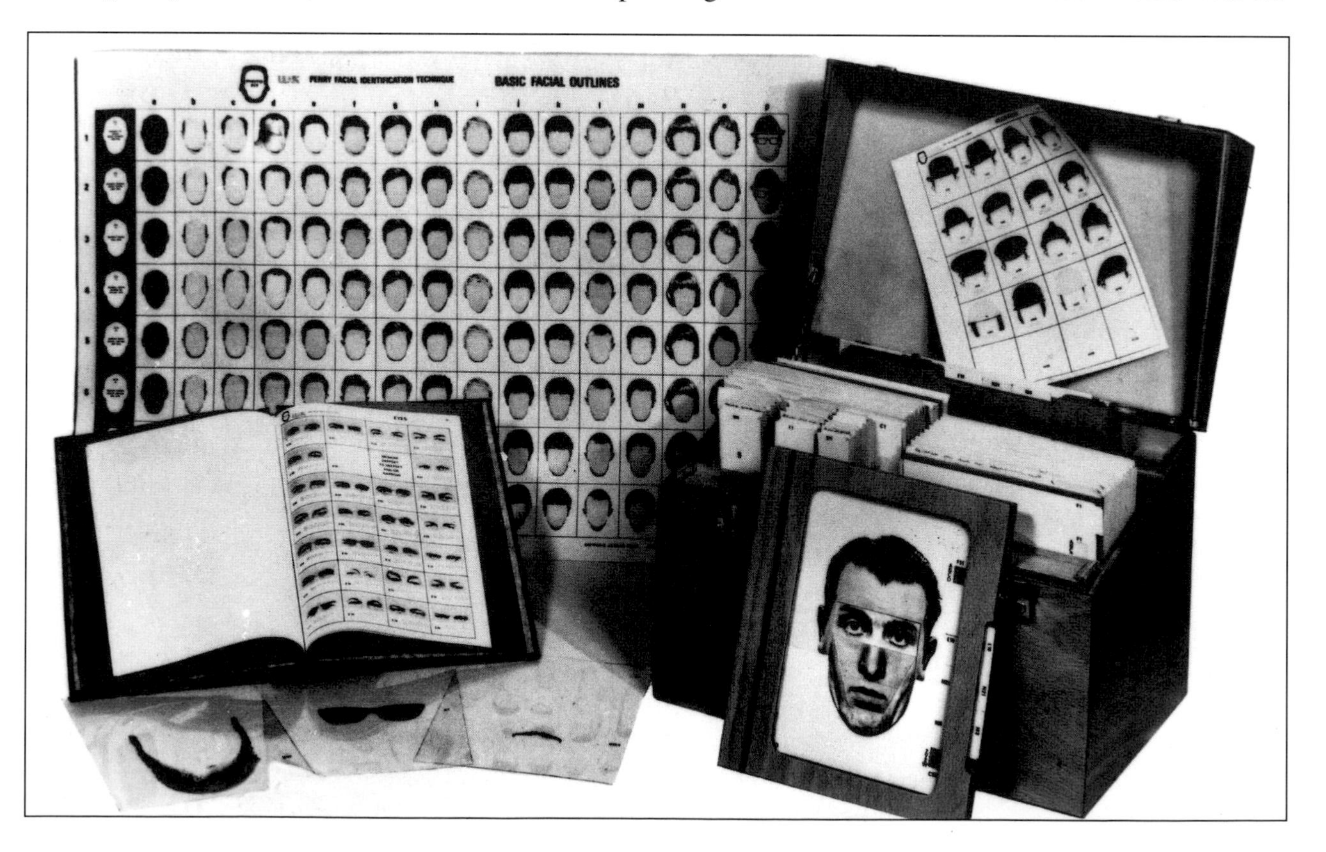

The Penry PhotoFIT, introduced in Britain in 1971, was an important step forward in visual identification. The Front-view Basic Caucasian Kit, shown here, was capable of composing some five billion different faces.

He found that witnesses' descriptions were often contradictory and incomplete. He began to make rough sketches on transparent sheets – different eyes, noses, facial shapes – which could be overlaid to produce a composite portrait that might be recognized by his informants.

On his return to the United States, McDonald took his ideas to the Townsend Company of Santa Ana, California, who were sufficiently impressed to develop the project. After several years of consultation with local police forces, the company produced the first Identikit pack. It comprised 525 coded and numbered transparencies, each with a drawing of a single feature, including 102 pairs of eyes, 32 noses, 33 pairs of lips, 52 chins, and 25 different moustaches and beards. In effect, this was the pictorial equivalent of Bertillon's *portrait parlé*. There were no ears, because, as McDonald pointed out: "Many victims of crime are usually facing the criminal at the all-important moment, and never see his ears properly. Special marks like big or deformed ears, or scars and moles, can be drawn in afterward on the slides with a wax pencil."

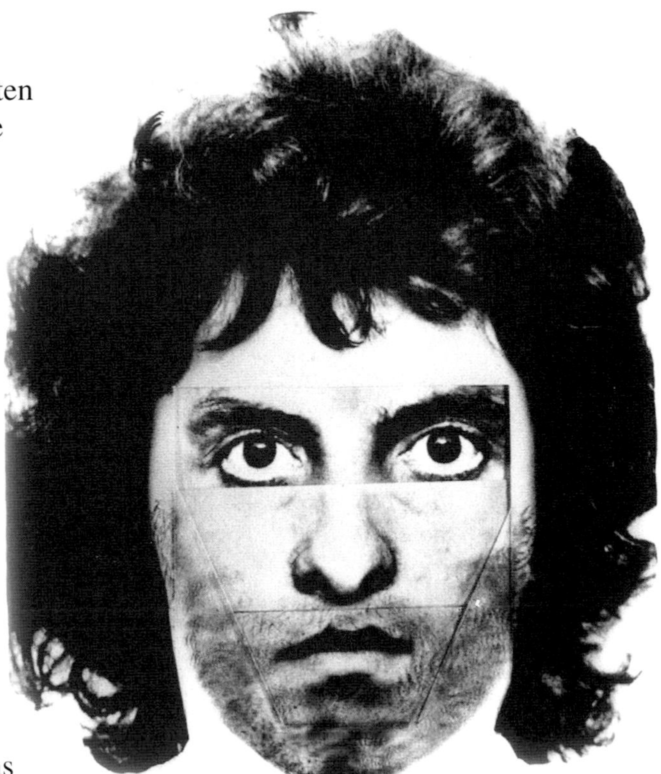

A PhotoFIT impression of a man wanted for the murder of an elderly woman found battered to death at her home in Tonbridge, Kent. The likeness was assembled from descriptions provided by several witnesses who had seen him visting the woman's home on a number of occasions.

McDonald claimed that it was possible to compile up to 62 billion different combinations of features. The coded system had another great advantage: at a time before fax machines were in use, an Identikit composite could be transmitted to another location as a set of the number and letter codes of the individual transparencies.

By 1960, the Identikit system was being used by many of the world's police forces. Although it could claim many successes, there were also a number of disturbing failures.

The introduction in 1971 of the Penry Facial Identification Kit – PhotoFIT – was a welcome improvement. Photographer Jacques Penry claimed that his system was not a development of Identikit; the idea had first come to him in 1938, when he was selecting photos to illustrate his book *Character from the Face*. However, 30 years passed before he approached the British Home Office Police Research and Development Branch, and was given a contract to produce his first kit.

PhotoFIT is similar to Identikit, but employs photographic elements rather than drawn images. Penry's first "Front-view Basic Caucasian Kit", produced in 1969, was capable of composing 5 billion different faces, and in 1970 the Afro-Asian Supplement added a further 500 million. Other supplements

CRIME FILE: Anton Fähndrich

A leading handwriting expert was able to provide a detailed profile of the young man responsible for a series of bombings in Switzerland. When police found the man, he matched the description closely.

On the evening of June 30, 1962, a bomb exploded in the elevator of a restaurant in Lucerne, Switzerland. Within twenty-four hours, there were four more incidents: in two more restaurant lifts, a cellar, and beneath a car. In total, five people were injured, one seriously, and 100,000 Swiss francs worth of damage was caused. Soon after, two further explosions followed.

When the police examined the debris of the bombs, they discovered that the detonators were manufactured items. Inquiries led them to a firearms dealer who remembered selling them some days previously to a man who had signed his register with the name "Afled Späni", and given a false address.

The police called upon the services of Zurich's leading graphologist, M. Litsenow. He said that it was well-nigh impossible to determine a personality from a single name and address, but agreed to try. The signature, he said, was obviously false, because it lacked all spontaneity. It was written by a simple person of only average intelligence, aged between 20 and 40, but probably nearer 20. He had been a poor pupil at school, and was an unstable personality.

Although he had constructed the bombs, the bomber was not a technician – he had no profession. His evident instability and feelings of inferiority made it unlikely that he worked in a shop, in contact with the public. If he was an agricultural labourer, he would not have known the whereabouts of the restaurants in Lucerne, and factories were rare in the city. He was probably a casual worker: a storeman, or something like that.

In view of his inferiority complex, said M. Litsenow, the bomber's motives derived from a need to feel important. At the same time, he would be outwardly conventional. His handwriting indicated that he was robust, even athletic, and probably good at sports.

He almost certainly knew someone called Alfred, and the address he had given had some connection with his life. His parents had probably been alcoholics, or separated. M. Litsenow suggested that the police should consult the social services. Also, the man might well have been in trouble with the police before, over a minor incident.

Armed with these indications, the police rounded up half a dozen young men, and eliminated all the suspects but one. His name was Anton Fähndrich, 20 years old and a storeman. Discreetly dressed, he had swept-back brown hair and a small moustache, and lodged in a church hostel, where he was very popular.

Comparison of the suspect's handwriting with the false signature revealed many similar characteristics – and the address was one where he had previously been a casual worker. Furthermore he had recently won two boxing championships. Questioned, Fähndrich said that his parents had separated, after being arrested several times for drunken behaviour. At first he denied any responsibility for the bombings, but eventually confessed that they were from "a need to revenge himself on society". It was the noise of the bombs, the shouts of the public, and the sound of police sirens that had excited him.

CASE CLOSED

followed, the Female Supplement being produced in 1974. The complete Caucasian Basic Kit alone comprised 204 foreheads and hairstyles; 96 pairs of eyes; 89 noses; 101 mouths; 74 chin and cheek sections; and various "accessories" such as headwear, moustaches and beards, spectacles, age lines, and ears.

Both Identikit and PhotoFIT have now been largely replaced by a system of computer graphics – Video-Fit. A computer can store a vast number of photographic elements, each of which can be manipulated to change its relative dimensions. The image can be rotated or tilted to provide a fully three-dimensional representation, and colour and texture can be changed as necessary.

The newest facial identification program is FaceIt. This classifies faces by just 12 variable characteristics, such as the length of the nose, the distance between the eyes, and the cheekbone structure.

Graphology

A person's handwriting is as characteristic and individual as a fingerprint, and as impossible to change. People who suffer injury and become incapable of writing with one hand are often forced to write with the other – and gradually exactly the same characteristics emerge that were present in their original writing. Graphologists – or, as the police prefer to call them when their assistance is sought in a criminal inquiry, "handwriting experts" – claim to be able to identify these characteristics. Moreover, they claim that they can detect innate personality traits that can indicate a potential or actual criminal.

Whether a person is deliberately trying to disguise his or her handwriting, or is affected by physical or emotional circumstances, a host of tell-tale details will allow an expert eye to identify the writer. One's signature, for instance, although never identical from one occasion to the next, remains fundamentally the same – two signatures that were exactly identical would immediately arouse suspicion of forgery.

The analysis and comparison of handwriting is a long and complex business. The principles were first expounded in 19th-century France, and subsequently developed, principally in Germany and Switzerland. Graphologists begin by dividing the writing into three zones. Young children still

RICHARD HAUPTMANN IV

RICHARD HAUPTMANN

SAME SIGNATURE RECONSTRUCTED FROM LETTERS CUT OUT OF KIDNAP NOTE

MR. CHAS. LINBERG,
YOUR BABY IS SAFE BUT HE IS NOT
USING NO MEDICINES. HE IS EATING
PORK CHOPS PORK AND BEANS JUST WHAT
WE EAT. JUST FOLOW OUR DIRECTION
AND HAVE ONE HUNDRED THOUSEND
BUCKS READY IN VERY SHORT TIME
THATS JUST WHAT WE NEED

Evidence produced in the trial of Bruno Hauptmann for the kidnapping and murder of Charles Lindbergh's baby son in 1932. Similarities between the formation of some of the letters, and those in a 1934 application by Hauptmann for a passenger vehicle registration, have been circled.

Research in the development of a computerized handwriting recognition system for use in forensic science. The system can establish that two letters were written by the same person, and can also provide clues about the nature of an individual's personality.

often learn to write in exercise books where these three zones are defined for them within four horizontal rulings. As they grow older, some continue to write in the same way, but most choose to express their personalities in deviations from the standard form. Graphologists believe that these deviations are sure indicators of personality. The upper zone, they say, is the area of intellectual and spiritual qualities, ambition and idealism. The middle zone represents the individual's likes, dislikes, rationality and adaptability to everyday social life. The lower zone reveals the instinctive and subconscious urges, together with the sexuality and materialistic interests of the writer.

Next, the graphologist analyzes the slant of the handwriting, then the formation of letters, both individually and in groups. Large capitals, for example, are said to suggest a generous personality with a developed need for self-expression and attention. Small capitals reveal feelings of inferiority and a desire for a quiet life. Tall narrow capitals indicate a strong but repressed personality; an inability to make friends, inhibition, and consequent frustration. Sharply angled capitals reveal aggression and obstinacy, with a lack of adaptability; whereas rounded capitals indicate an affectionate and humorous nature – if they are wide and fat, they are a sign of easy-going laziness. "Enrolled" capitals, in which the ends of the strokes are scrolled in on themselves, are held to be a sure sign of deceitfulness.

One of the most revealing of all is the capital I: in handwriting it directly represents the ego. A small I, for instance, reveals a lack of self-confidence; while one that is flamboyant and exaggerated is a sure sign of someone who wants to be the centre of attraction. An I that is a simple downward stroke suggests a personality that is self-assured, intelligent and well balanced; but an I that swings or leans to the left, whatever its form, indicates an inability to enjoy life, possibly guilt about some past event, and a propensity to deceive.

These are just a few examples of the criteria employed by the graphologist.

Psychological Profiling

Cesare Lombroso's theories concerning the physical types of criminals have long been discarded, but it is only within the last 50 years that criminologists have seriously turned their attention to criminal psychology. The first full study was of Peter Kurten, the so-called "Vampire of Düsseldorf", written by Professor Karl Berg in 1930, but little more work was done until Dr. James Brussel made his remarkable assessment of the "Mad Bomber of New York" in 1957.

From the 1950s onward, criminologists took an increased interest in what became known as "psychological profiling". The theory received a serious setback in 1964, when a team of psychiatrists, of which Dr. Brussel was a member, attempted to develop a profile of the Boston Strangler. They concluded – although it must be said that Dr. Brussel disagreed – that there were two perpetrators: one a man who lived alone, probably a schoolteacher; and the other a homosexual with a hatred of women. When Albert DeSalvo was finally tracked down, however, he was found to be the only perpetrator – and he was an over-sexed married man with children.

In the FBI Academy at Quantico, Howard Teten took up the study of psychological profiling in 1969. He gathered much valuable advice from Dr. Brussel, and in 1972 he was joined by Pat Mullany. The two men began to build up a file of interviews with serial killers on tape, using a computer database to search for similar patterns of thinking. Together, they established the FBI's Behavioral Science Unit at Quantico, where they added Robert Ressler to the team, early in 1974.

The fledgling Unit soon had an opportunity to test their research. In June 1973, seven-year-old Susan Jaeger was abducted from a tent in which she was camping with her family near Bozeman, Montana. Teten and Mullany put together a preliminary profile: the perpetrator was a young white male who lived in the area, a loner who had come across the family camp while walking

CRIME FILE:

George Metesky

The identification of the "Mad Bomber of New York" was the first major success in psychological profiling. The technique now plays an essential part in the FBI's tracking-down of violent criminals.

Safely behind bars in Waterbury jail, the "Mad Bomber" George Metesky smiles for photographers after his arrest, which brought to an end a bombing campaign that had lasted, intermittently, for 16 years.

In November 1940, an unexploded small bomb was found on a window sill of Consolidated Edison, the electricity supply company for New York. With it was a note, written in capitals: "Con Edison Crooks, This Is for You!" A similar undetonated bomb was discovered in the street 10 months later. When Japan attacked at Pearl Harbor in December 1941, the police received another note, posted in Westchester County, New York. It read: "I will make no more bomb units for the duration of the war – My patriotic feelings have made me decide this – I will bring the Con Edison to justice later – They will pay for their dastardly deeds. F.P."

Over the next five years, similar notes were received by Con-Ed, newspapers, hotels and department stores. Then there were no more, and the police decided that "F.P." had given up his campaign, or died. Then, on March 25, 1950, another unexploded bomb was found in Grand Central Station.

All the bombs had been carefully made, and it seemed that the "Mad Bomber", as he became known, did not mean them to detonate. The next, however, planted in a telephone booth, did explode, and letters to newspapers threatened more – "To Serve Justice". During the next four years twelve further bombs exploded, and six were planted in 1955, two of which failed to detonate. The bombs were becoming increasingly destructive, although only four people had been slightly injured, and the bomber was obviously becoming angrier. In a letter to the *Herald Tribune* he claimed; "These bombings will continue until the Con Edison is brought to justice."

On December 2, 1956, a bomb exploded in the Paramount Theater in Brooklyn, injuring six people, three of them seriously. Inspector Howard E. Finney of the New York Police Crime Laboratory decided to take what was then a most unusual course, and consulted psychiatrist Dr. James A. Brussel.

Dr. Brussel provided Finney with a remarkably detailed assessment. In his opinion, the Mad Bomber was a male, an acute paranoiac, aged around 50. He was well built, clean-shaven, and meticulous in appearance. He was a loner, unmarried, but probably lived with an older female relative. English was not his native tongue, and he was an immigrant or the son of one, probably a Slav or a Pole. "And when you catch him, he'll be wearing a double-breasted suit. Buttoned."

The police arranged for the publication of a summary of Dr. Brussel's conclusions, and were rewarded by a letter to the *Journal American* in which the bomber stated that he had been injured at a Con-Ed plant, was permanently disabled, and had received no compensation. While they were still going through Con-Ed's personnel records, the man himself provided the final clue in another letter: "I was injured on job at Con Edison plant – September 5th 1931."

The records provided only one name that fitted: George Metesky, born 1904, the son of a Polish immigrant. There was a large Polish community around Bridgeport, Connecticut, and Westchester, where the letters had been posted, lay between Bridgeport and New York City. In Waterbury, near Bridgeport, detectives called at the home of Metesky, where he lived with two elderly half-sisters. He was well built but, as it was late at night, he was wearing a dressing gown. The detectives told him to get dressed, and when he re-appeared he was wearing a shirt and tie, and a blue double-breasted suit – buttoned.

Metesky was found to be unfit to stand trial, and was committed to an asylum for life. The letters "F.P.", he said, stood for "Fair Play".

Some of the bomb-making materials discovered in the garage of Metesky's home on the morning following his arrest.

Dr James A. Brussel, the New York psychologist who achieved a remarkable success with a description of the "Mad Bomber" George Metesky, with a copy of his Casebook of a Crime Psychiatrist.

at night. They concluded that the girl was probably dead.

The Bozeman FBI agent, Peter Dunbar, had a suspect who fitted this description: 23-year-old Vietnam veteran David Meierhofer, but there was no material evidence to connect him with the abduction. Then, in January 1974, an 18-year-old girl who had rejected Meierhofer also went missing. He was again a suspect, but he volunteered to take polygraph and "truth serum" tests, and passed both.

However, the Quantico team had further information to work on, and their refined profile definitely fitted Meierhofer. They knew that many psychopaths are able to separate the personality responsible for crimes from their in-control selves, and so defeat the polygraph. Teten and Mullany thought that the suspect might well be the type who telephones the relatives of his victims, to relive the excitement of the crime. Dunbar asked Mr. and Mrs. Jaeger to keep a tape recorder beside their telephone.

On the anniversary of their daughter's abduction, Mrs. Jaeger received a call from a man, who said he had taken her to Europe, and was giving her a better life than her parents could afford. An FBI voice analyst concluded that the voice on the tape matched Meierhofer's, but this evidence was not considered sufficient in Montana to obtain a warrant to search his home. Mullany arranged for Mrs. Jaeger to confront Meierhofer at his lawyer's office; during the meeting he was cool and collected, but shortly after Mrs. Jaeger returned home she received a call from a "Mr. Travis of Salt Lake City", who said that it was he who had abducted Susan. Before he could say more, Mrs. Jaeger interrupted him. "Well, hello David," she said.

Dunbar was now able to obtain a search warrant, and the remains of the two missing girls were found in Meierhofer's home. He confessed to the two murders, as well as the unsolved killing of a local boy. He was arrested, and the following day he hanged himself in his cell.

In the United States, the FBI was not alone in its interest in the psychology of the serial killer. As early as 1957, a Los Angeles detective named Pierce Brooks was put in charge of the investigation of the apparently unrelated rapes and murders of two young women. He concluded that the same man was responsible for both, and spent weeks going through newspaper files, looking for other murders that matched the killer's MO. When Harvey Glattman was eventually apprehended for the crimes, Brooks obtained a detailed confession from him, and this forms one of the earliest documents of a serial killer's mind-set.

In July 1983 Brooks – now a consultant with 35 years of police work behind him – appeared before a Senate sub-committee in Washington DC. Together with Roger Depue, then head of the Behavioral Science Unit, he proposed the setting-up of a Violent Criminal Apprehension Program (VICAP). A few months later, President Reagan announced the establishment of a National Center for the Analysis of Violent Crime (NCAVC).

Hitherto, working out of Quantico, the FBI had relied largely upon scene of crime photographs to provide a detailed description of a crime and its surroundings. Now they introduced the VICAP Crime Analysis Report, which was distributed to all of their 59 field divisions. The questions on the form total 189, from "case administration data" to "crime classification" (including likely related crimes), details of the crime and its victims, MO, autopsy data and forensic evidence. Using this form, investigating officers can request the NCAVC to compare a crime with hundreds of others held on a computer database.

Since 1990, the Behavioral Science Unit has been renamed Behavioral Science Services, with an Investigative Support Unit running a Criminal Investigation Analysis Program. In one year alone, it examined 793 cases, 290 of which fell directly within FBI jurisdiction.

In other countries, the value of psychological profiling is recognized, although it is not generally organized on the same scale.

Between 1982 and 1986, London detectives were hunting a man who committed at least 30 rapes and 3 murders. In July 1985 he attacked three women in one night. The only clues were in his MO: both rapes and murders occurred close to railway stations, and all three murder victims were strangled in the same way, but the locations of the attacks were spread all over London and the Home Counties.

Using information given to him by the police, David Canter, Professor of Applied Psychology at the University of Surrey, drew up a profile of the murderous rapist. He lived in the Kilburn–Cricklewood area of north London; was a semi-skilled worker, with a knowledge of the London railway system, and had a job that did not bring him into much contact with the public; was married but childless, and had a turbulent relationship with his wife; and had one or two close male friends.

CRIME FILE:

Richard Trenton Chase

Sitting at his desk in Quantico, FBI expert Robert Ressler was able to provide a profile of a psychopathic killer in California.

On January 23, 1978, Sacramento truck-driver David Wallin arrived home to find his 22-year-old wife Theresa butchered in the bedroom. A yoghurt container beside the body showed signs of having been used to drink the blood of the eviscerated victim, and several body parts were missing. There appeared to be no motive for the crime.

As it happened, Robert Ressler, of the FBI Behavioral Science Unit, was due to visit the West Coast. Before leaving, he wrote out a preliminary profile:

> White male, aged 25–27 years; thin, undernourished appearance. Residence will be extremely slovenly and unkempt and evidence of the crime will be found at the residence. History of mental illness, and will have been involved in the use of drugs. Will be a loner who does not associate with either males or females, and will probably spend a great deal of time in his own home, where he lives alone. Unemployed. Possibly receives some form of disability money. If residing with anyone, it would be with his parents; however, this is unlikely. No prior military record; high school or college dropout. Probably suffering from one or more forms of paranoid psychosis.

Even before Ressler reached California, the killer struck again. On January 26, three bodies were discovered in a house within a mile of the Wallin murder. Evelyn Miroth, 36 years old, had been mutilated in a manner even worse than Theresa Wallin; her 6-year old son Jason, and Daniel Meredith, a family friend, had been shot; and Evelyn's baby nephew was missing. Judging by the amount of blood in the baby's crib, police were sure that he, too, was dead. Meredith's station wagon had been taken, but was found abandoned not far away.

"With a mounting sense of urgency and the certainty that…this man would kill again", Ressler added more details to his profile: "single, living alone…one-half to one mile from the abandoned station wagon". He believed the man had previously committed "fetish" burglaries in the area, stealing such things as women's clothing, rather than anything of value.

Following Ressler's recommendation, police concentrated their inquiries in the small area. They found a witness who had spoken to a young man she had known at high school, Richard Trenton Chase, and had been shocked at his appearance; "dishevelled, cadaverously thin, bloody sweatshirt, yellowed crust around his mouth, sunken eyes".

Police staked out Chase's home, lured him out, and arrested him in possession of a .22 revolver and Meredith's wallet. In his truck they found a 12-inch (30-centimetre) butcher's knife, and blood-caked rubber boots. His home was filthy, and several containers in the refrigerator held body parts and human brain tissue. A calendar was marked with the word "Today" on the dates of the murders, and the same word was written on 44 more dates throughout 1978.

After Chase was caught, computer searches and interviews revealed that Ressler had been phenomenally accurate in his assessment of the killer.

CASE CLOSED

The police had no less than 1999 suspects on computer. When Canter's profile was compared with the database, John Francis Duffy's name came out top, and when forensic evidence identified him as the guilty man, the profile was found to be correct in 13 out of 17 points.

Another British psychologist, Paul Britton, has been involved in a number of police investigations, including an unusual one that had an unexpected outcome.

In August 1988, Pedigree Pet Foods received a letter claiming that cans of their food were being contaminated with poison, and the writer demanded the payment of £500,000 over five years. Britton was consulted: he gave his opinion that the threat was serious. The letter revealed that its writer was not psychotic, a male, intellectually average or above, and normally educated, although not to university level. Britton recommended that Pedigree should begin paying sums into the many savings accounts, under assumed names, that were listed by the blackmailer; in due course, a pattern of withdrawals from cash dispensers might emerge.

A week after the first payments were lodged, cash withdrawals began. They took place almost nightly, but at locations all over the British Isles. After plotting these on a map, and reasoning that the blackmailer worked from a centre, Britton concluded that he lived near Hornchurch, on the east side of London. He was not a young man, because he had planned his campaign patiently, and – because he was free to travel – was probably retired.

The police had kept the case secret, mounting undercover surveillance on randomly selected cash machines throughout the country, but with no success. Then a newspaper revealed what was happening, and the blackmailer changed his target – to Heinz. There was no doubt that he was serious: jars of baby food were soon found contaminated with caustic soda or broken razor blades. Heinz were unwilling to follow the example of Pedigree, but eventually agreed to begin payments into building society accounts.

In a private talk with the police, Britton made his most sensational pronouncement: he believed the blackmailer to be a former police officer. The man showed a great deal of familiarity with police investigative procedure – as well as a detailed knowledge of a previous similar case – and even seemed to know where surveillance was mounted. "He's tenacious," said Britton, "but hasn't done well in the police service. He thought he was going places, but his career stalled, and he blames his superiors...he's demonstrating just how good he could have been."

Finally, on October 20, 1989, a surveillance team arrested a man approaching a cash dispenser in north London. His name was Rodney Witchelow, 43 years old, living in Hornchurch. A former detective in the East London Regional Crime Squad, he had retired on medical grounds in October 1988, but had kept up friendships with workmates – and on one occasion had actually sat in a police car engaged in the surveillance operation.

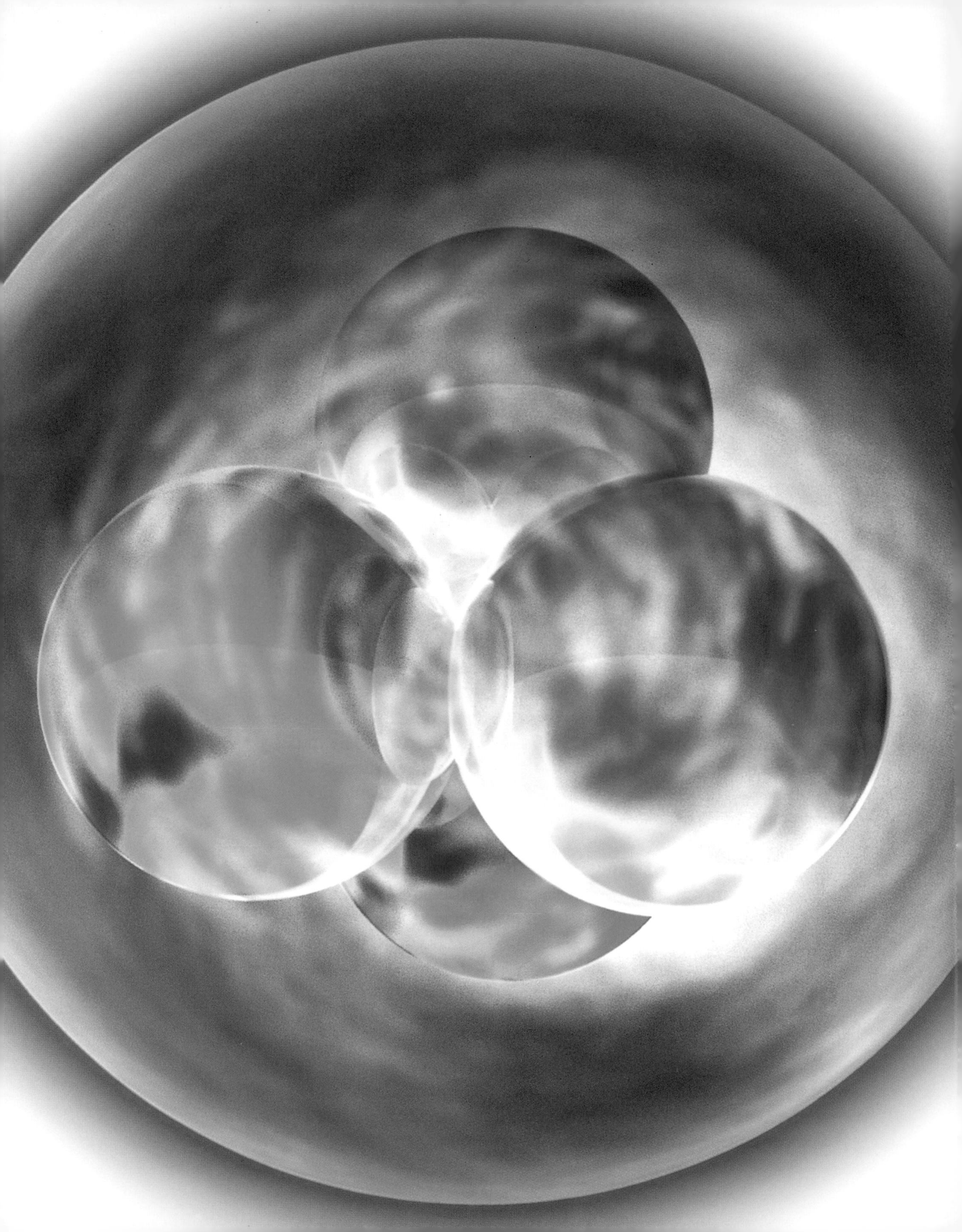

The Forensic Hardware

A GREAT DEAL OF FORENSIC INVESTIGATION OF A CRIME is concerned with trace evidence: blood, semen and sweat; explosive residues; poisons; hairs and fibres; fragments of paint, glass, paper and ink, earth and sand.

These traces are seldom, if ever, uncontaminated with other material, and laboratory examination uses specialized techniques and equipment to separate and identify each of their separate components. A brief explanation of the principal analytical methods employed may help the reader to understand the scientific principles involved, and to appreciate the sophistication of current forensic investigation.

Chromatography

In all its various forms, chromatography is an elegant way of separating and identifying individual chemical compounds from a mixture. It was given its name by the original developer of the technique, the Russian botanist Mikhail Tsvet in 1906, because he used it to separate plant pigments.

In principle, all chromatography involves two "phases": a stationary phase – a material that will adsorb the components of the mixture – and a mobile phase in which all the components are soluble. The separation of the components depends on competition for molecules between the stationary phase and the mobile phase. As the mobile phase moves through the stationary phase, the different components are adsorbed at different rates, and gradually become separated.

A computer artwork of the nucleus of a helium atom, or alpha particle, given off during radioactive decay. The nucleus consists of two positively charged protons (red) and two neutral neutrons (green) surrounded by a quantum cloud of gluons.

Tsvet's technique was very simple. He placed his sample, a solution of various plant pigments in alcohol, at the top of a glass tube containing a column of aluminium oxide. As he added more alcohol, the solution moved down the tube, leaving the pigments gradually separated into bands. Finally, the individual pigments were washed out as separate fractions in alcohol solution.

In a modern refinement, the emerging fractions are detected by an optical monitoring device. This measures the absorption of ultraviolet light (see Spectrometry) as the solvent flows from the bottom of the column, and records this as a pen trace on a slowly moving band of paper.

Other refinements of chromatography include paper chromatography and thin-layer chromatography, both of which allow the direct identification of components, without further analysis. Paper chromatography uses a sheet of filter paper as the stationary phase, thin-layer chromatography a thin film of aluminium oxide, or silica gel, on a glass plate. The lower end of the stationary phase is dipped into a suitable solvent, which moves upward by capillary action.

Mikhail Tsvet, the Russian botanist who first developed the technique of chromatography.

A spot of the sample to be analyzed is placed at the bottom, and control samples of known substances are placed alongside it. When the solvent has reached the top, the paper or plate is dried, and the separated spots are located by spraying with a suitable reagent, or illuminating with ultraviolet light. If a spot from the unknown sample has moved the same distance as a known substance, this identifies that particular component of the mixture.

Gas chromatography is used for the separation of mixtures of both liquids and gases. Here, the stationary phase is coated on fine clay or glass beads, which are packed in a steel tube, and the mobile phase is the actual mixture to be analyzed, which is blown through the apparatus by an inert gas such as nitrogen. Liquids must be heated above their boiling point, and the steel tube must also be heated. Various types of detector measure the emerging fractions.

Solid samples, such as fibres and paints, can also be analyzed by a development of this method, pyrolysis gas chromatography. The sample is heated to a temperature at which it decomposes into gaseous components. The resultant trace from the detecting device is usually sufficiently characteristic to be compared with that from known materials, and so identified.

Electrophoresis

The basic principles of electrophoresis have been outlined in "Written in Blood" and "DNA Fingerprinting". In essence, this technique is similar to chromatography, in that it depends upon the different rates at which molecules move through a stationary phase. In electrophoresis, this movement – "migration" – is produced by a small direct electric current across the stationary phase.

Ordinary filter paper or a synthetic membrane can serve as the stationary phase, but gels such as starch or silica, coated in a thin layer on a glass plate, are most commonly used.

Although electrophoresis can be used to analyze a wide range of molecules, it is particularly valuable in the separation of proteins.

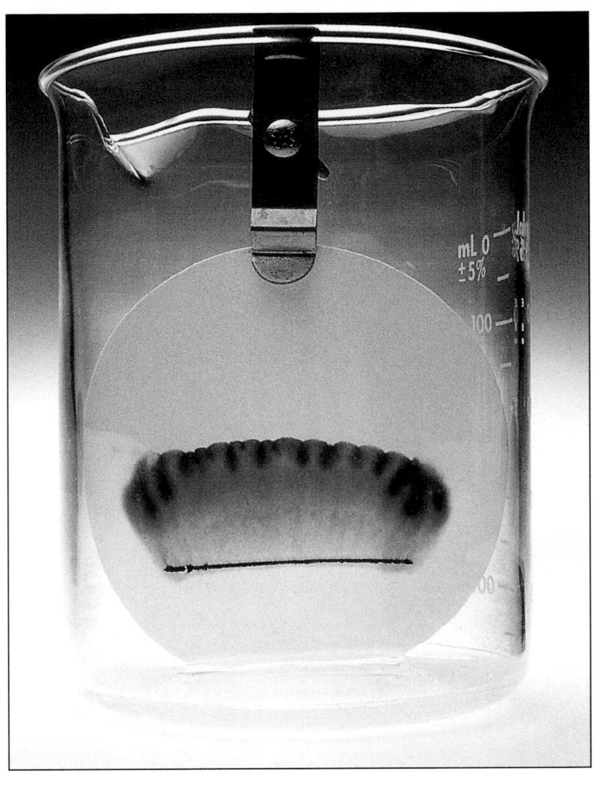

Paper chromatography demonstrated as coloured bands on a piece of filter paper dipped into a solvent. The mixture to be separated is a line of black ink drawn on the paper. As the solvent is drawn up the paper, it carries the components of the dye with it at different speeds – hence the different coloured bands.

Mass Spectrometry

The mass spectrometer is one of the most sophisticated pieces of equipment used in forensic chemical analysis. It can analyze an organic compound – one that has a structural "skeleton" made up of carbon atoms – in terms of its constituent parts. When a mixture of organic compounds is separated by gas chromatography (see above), many components of the mixture are often present in quantities that are too small to be confirmed by conventional chemical analysis. This is where mass spectrometry can be vitally important.

The sample under analysis is bombarded with electrons produced by a heated wire cathode. This breaks the molecules in the sample into fragments, each of which is electrically charged. The fragments then pass into the spectrometer through an electric field, which accelerates them. After this, they enter a magnetic field, which deflects them from their straight path into a circular one. The radius of this circular path varies according to the mass of the individual fragments: heavy fragments follow a path with a large radius, while lighter fragments are deflected into a path with a smaller radius. The radius of the path also depends on the strength of the magnetic field: as this is increased, the radius of the heavier fragments' path decreases.

The mass spectrometer is curved, and has a narrow slit at the end, with a detector on the other side. When the magnetic field is weak, only the lightest fragments are deflected enough to pass through this slit. As the magnetic field increases, heavier particles are deflected, and pass through the slit. If the detector is moved across the slit as the magnetic field increases, the result is a spectrum of the different fragments. The position of each fragment in this spectrum is a measure of its mass, and the intensity of each will be a measure of the respective proportion of each fragment. A knowledge of

Much of the equipment now available to the forensic scientist is automated, doing away with the need for laborious sequential analysis. Here a gas chromatography unit (left) is connected directly to a mass spectrometer. The technician is studying the computer read-out.

chemical structure makes it easy to identify the compound in the sample. In practice, the detector is linked to a computer, which provides a very rapid analysis.

The Microscope

A simple microscope uses light reflected through the sample being analyzed, or from its surface, and magnified by a system of lenses. Even this can be of great value in identifying trace evidence. The comparison microscope, in which two samples are placed side by side and viewed through a single eyepiece, has been described in "The Speeding Bullet".

The phase-contrast microscope is very useful in fibre identification, and for examining biological tissues. In effect, this causes some of the light waves passing through the sample to be "out of step" with others. As a result, parts of the internal structure of the sample – which would otherwise appear transparent – are seen as relatively sharp shadows.

The magnification, and therefore the resolution, in an ordinary optical microscope is limited by the wavelength of the visible light: structures that are smaller than the wavelength cannot be seen. The electron microscope was developed to overcome this problem. Although electrons are considered as tiny particles, they also behave as waves, and their wavelength is very much shorter than that of visible light.

The limit of magnification in the best optical microscope is about 2000 times. Electron microscopes are fundamentally of two types. The transmission microscope, in which a beam of electrons passes through a very thin sample, can provide a photographic image magnified more than one million times. The scanning microscope, which is the type of value in forensic examination, reflects electrons from the surface of the sample. Its effective magnification is some 150,000 times.

Neutron Activation

Radioactive elements emit three kinds of radiation: alpha particles (helium nuclei), beta particles (electrons), and gamma rays (such as X-rays). Other elements can be made radioactive by bombarding them with neutrons in the core of a nuclear reactor. The gamma rays that are then emitted by these elements can be detected, and their energy measured: each element emits gamma rays of a characteristic energy level. The method can be used to identify extremely small traces of elements, and their proportions, in metals, paint, glass, fibres, and other materials.

Refractometry

The laboratory measurement of the refractive index of a material such as glass is described in "Fragments of Evidence". Instruments known as refractometers are also used for refractive index measurement, particularly in the case of liquids.

Refraction occurs because the speed of light through a given material is less than that in a vacuum (or, for practical purposes, air). The refractive index is the ratio of these two speeds. A coloured spectrum is produced by a glass prism, or in a rainbow, because the refractive index varies with the wavelength of light. Refractometers therefore use light of a single wavelength, usually yellow sodium light.

A typical example is the Pulfrich refractometer. This consists of a polished glass block, with a small hollow on the upper surface for liquids. A beam of sodium light is aimed through the block from below, and measurement of the angle at which it emerges makes it possible to calculate the refractive index of the liquid.

Spectrometry

The separation of light into its component wavelengths of electromagnetic radiation results in a spectrum. There is a range of wavelengths visible to the human eye: the longest is red, and the shortest is violet. In addition, there is a range of longer wavelength than red, the infra-red; and of shorter wavelength than violet, the ultraviolet. Spectrometers are designed to detect all these different wavelengths.

When a beam of electromagnetic radiation shines through a material, some of the wavelengths are absorbed. This, for example, explains why a material appears blue in colour: it has absorbed the red wavelengths. The specific wavelengths absorbed are characteristic of the molecules in the material, and so the method can be used to identify a mixture of components.

Emission spectrographs make use of the fact that, when elements are heated to a high temperature, they will emit light of characteristic wavelengths. They are of particular importance in the analysis of glass, paint and metals. The sample is heated in a carbon arc, by means of a laser, or by electron bombardment, as in the mass spectrometer, and the emitted light is focused through a glass prism to produce a spectrum. This is not a continuous spectrum, but consists of a series of lines of different colours, each representing a specific wavelength. Because glass absorbs ultraviolet light, an alternative device, known as a diffraction grating, replaces the prism when a broader spectrum is required.

The absorption spectrometer employs the opposite principle. Elements vaporized in a flame will absorb specific wavelengths. A radiation source, shone through the flame and then through a diffraction grating, will reveal the wavelengths absorbed as a series of dark lines in the spectrum.

Both emission and absorption spectroscopy are destructive of evidence, but only very small samples are required.

Bibliography

Baden, Michael: *Unnatural Death*; Ballantine, New York 1989

Bellemare, Pierre: *Les Nouveaux Dossiers Extraordinaires*; Editions de la Seine, Paris 1990

Block, Eugene B.: *Fingerprinting*; Franklin Watts, New York 1969

Britton, Paul: *The Jigsaw Man*; Bantam, London 1997

Campbell, Marjorie Freeman: *A Century of Crime*; McClelland & Stewart, Toronto 1970

Canter, David: *Criminal Shadows*; HarperCollins, London 1994

Cooper, Paulette: *The Medical Detectives*; McKay, New York 1973

Douglas, John and Mark Olshaker: *Mindhunter*; Heinemann, London 1996

____.*Journey into Darkness*; Heinemann, London 1997

____.*Obsession*; Simon & Schuster, London 1998

Ellenhorn, Matthew J.: *Medical Toxicology*; Williams & Wilkins, Baltimore 1997

Evans, Colin: *The Casebook of Forensic Detection*; Wiley, New York 1996

Federal Bureau of Investigation: *The Science of Fingerprints*; n.d.

Fisher, Barry A.J.: *Techniques of Crime Scene Investigation*; Elsevier, New York 1992

Fisher, David: *Hard Evidence*; Simon & Schuster, New York 1995

Fletcher, Tony: *Memories of Murder*; Weidenfeld & Nicholson, London 1986

Gaute, J.H.H., and Robin Odell: *The New Murderers' Who's Who*; Headline, London 1989

____. *Murder 'Whatdunit'*; Harrap, London 1982

Glaister, John: *Medical Jurisprudence & Toxicology*; Livingstone, Edinburgh (many editions)

Hastings, Macdonald: *The Other Mr Churchill*; Harrap, London 1963

Holme, David J., and Hazel Peck: *Analytical Biochemistry*; John Wiley & Sons, New York 1993

Holmes, Ronald M.: *Profiling Violent Crimes*; Sage Publications, Newbury Park California 1989

Joyce, Christopher, and Eric Stover: *Witnesses from the Grave*; Bloomsbury, London 1991

Kind, Stuart: *Scientific Investigation of Crime*; Forensic Science Services, Harrogate 1987

Kind, Stuart, and Michael Overman: *Science against Crime*; Aldus, London 1972

Knight, Bernard: *Simpson's Forensic Medicine*; Edward Arnold, London 1991

Lambourne, Gerald: *The Fingerprint Story*; Harrap, London 1984

Lane, Brian: *Encyclopedia of Forensic Science*; Headline, London 1992

Lucas, Norman: *The Laboratory Detectives*; Arthur Barker, London 1971

McGrady, Mike: *Crime Scientists*; Lippincott, New York 1961

McLay, W.D.S. (ed): *Clinical Forensic Medicine*; Pinter, London & New York 1990

Marne, Patricia: *The Criminal Hand*; Sphere, London 1991

Marriner, Brian: *Forensic Clues to Murder*; Arrow Books, London 1991

Miller, Hugh: *Traces of Guilt*; BBC Books, London 1995

____. *Proclaimed in Blood*; Headline, London 1995

____. *Forensic Fingerprints*; Headline, London 1998

Nash, Jay Robert: *World Encyclopedia of 20th Century Murder*; Headline, London 1992

Olson, Kent R.: *Poisoning & Drug Overdose*; Prentice-Hall, New Jersey 1990

Paul, Philip: *Murder under the Microscope*; Macdonald, London 1990

Polson, C.J., D.J.Gee and B.Knight: *Essentials of Forensic Medicine*; Pergamon Press, Oxford 1985

Quigley, Christine: *The Corpse, a History*; McFarland, Jefferson NC 1996

Ragle, Larry: *Crime Scene*; Avon Books, New York 1995

Ressler, Robert K. and Tom Schachtman: *Whoever Fights Monsters*; Simon & Schuster, London 1992

Ressler, Robert K., John E. Douglas, Ann W. Burgess and Allen G. Burgess: *Crime Classification Manual*; Simon & Schuster, London 1993

Scott, Gini Graham: *Homicide*; Roxbury Park, Los Angeles 1998

Simpson, Keith: *Forty Years of Murder*; Harrap, London 1978

Smith, Sir Sydney: *Mostly Murder*; Harrap, London 1986

Smyth, Frank: *Cause of Death*; Orbis, London 1980

Stern, Chester: *Dr Iain West's Casebook*; Little, Brown, London 1996

Stockdale, R.E. (ed): *Science against Crime*; Marshall Cavendish, London 1982

Thompson, John: *Crime Scientist*; Harrap, London 1980

Timbrell, J.A.: *Introduction to Toxicology*; Taylor & Francis, London 1989

Tullett, Tom: *Clues to Murder*; Bodley Head, London 1986

Walls, H.J.: *Forensic Science*; Sweet & Maxwell, London 1968

Ward, Jenny: *Crimebusting*; Blandford, London 1998

Wecht, Cyril: *Cause of Death*; Virgin, London 1993

____.*Grave Secrets*; Onyx, New York 1998

Williams, Judy: *The Modern Sherlock Holmes*; Broadside, London 1992

Wilson, Colin: *Written in Blood*; Equation, Northampton 1989

Wynn, Douglas: *On Trial for Murder*; Pan Books, London 1996

Index

Picture Credits

AKG: 30, 112, 117, 164. **Amber Books**: front cover (br), 141. **Associated Press (AP)**: 28 (both), 51 (t), 62 (t), 90, 107, 159, 168, 169, 170, 230 (b), 231. **Neville Chadwick Photography**: 156, 157. **Corbis (Bettmann / UPI)**: back cover (bl & br), 38, 52 (both), 53, 61 (b), 65, 68, 77, 78-79 (all), 81, 86-87 (all), 89, 95, 97, 122, 124, 130, 175, 176 (both), 177, 178, 182 (both), 183, 196 (Reuters), 202, 220, 223, 224 (t), 226-227, 228, 229, 233, 234, 237, 240, 241, 242. **Corbis-Sygma**: back cover (bc), 191 (both), 192, 193, 198 (both), 199 (Oakland Tribune). **ESDA / GF Design**: 217. **Mary Evans Picture Library**: 29, 48, 51 (b), 55 (both), 58, 63 (t). **The Fairfax Photo Library (Sydney/Aust)** : 210. **Harrap Limited / Sir Sydney Smith**: 83 (t). **Hulton Getty**: 174. **Katz Pictures**: 67. **Kobal Collection**: 60. **Ben Martin/ (c) Time Inc./ Katz**: 214. **Richard Neave / Unit of Art in Medicine, Manchester University**: 88 (all). **Oxford Scientific Films**: 32-33, 36, 46-47, 54 (tl), 70-71. **PA Photos**: 13, 37, 57, 61 (tr), 101, 166, 194, 197, 230 (t), 235. **Pictor International**: front cover (bl & bc), 18, 135, 206-207, 208, 212, 215. **Popperfoto**: 10, 14, 15, 16 (both), 17, 24, 31, 35 (both), 222 (all). **Public Record Office Image Library**: 42-43 (all), 82, 83 (br & b), 98 (both). **Science Photo Library**: back cover (t), 8-9, 11, 12, 22-23, 34, 54 (tr), 62 (b), 63 (b), 69, 72 (all), 92-93, 103, 104-105, 106, 108, 109, 110-111, 118, 119 (both), 120, 123, 125, 126, 132, 136-137, 138, 139, 140, 144, 146-147, 148, 150, 153, 155, 158, 160-161, 162, 163, 165, 171 (both), 172-173, 186-187, 188, 189 (both), 190, 201, 203, 218-219, 221, 224 (b), 238, 246-247, 248, 249, 250. **The Wellcome Trust Medical Photographic Library**: 26, 40 (National Medical Slide Bank), 41 (National Medical Slide Bank), 45, 49, 54 (b), 84, 94, 96 (both), 99 (National Medical Slide Bank), 152 (National Medical Slide Bank), 180, 184 (National Medical Slide Bank). **US Department of Justice / Federal Bureau of Investigation**: 114. **Western Mail & Echo**: 74 (both), 75. **Western Media Publishing Ltd**: 142, 143 (both).